Knowledge and Conditionals

Knowledge and Conditionals

Essays on the Structure of Inquiry

Robert C. Stalnaker

OXFORD
UNIVERSITY PRESS

OXFORD
UNIVERSITY PRESS

Great Clarendon Street, Oxford, OX2 6DP,
United Kingdom

Oxford University Press is a department of the University of Oxford.
It furthers the University's objective of excellence in research, scholarship,
and education by publishing worldwide. Oxford is a registered trade mark of
Oxford University Press in the UK and in certain other countries

First Edition published in 2019

Published in the United States of America by Oxford University Press
198 Madison Avenue, New York, NY 10016, United States of America

British Library Cataloguing in Publication Data
Data available

Library of Congress Control Number: 2019933997

ISBN 978–0–19–881034–6

Contents

Acknowledgments

Nine of the papers collected here have been previously published, or in two cases, are forthcoming in other publications. I thank the editors and publishers for permission to reprint them here. The details about the sources are listed on page iv.

I have too many intellectual debts to acknowledge them all at this point, but let me mention some of the people that have affected my thinking on all the issues I discuss in the chapters of this book.

Many of the ideas in these papers were developed in seminars I gave at MIT and at Columbia University on conditionals and on topics in epistemology over the past five or six years. I was fortunate to have groups of very talented philosophers participating in those seminars whose critical and constructive contributions to the discussion helped me to understand the issues, and influenced my responses to them. These included Jessica Collins, Nilanjan Das, Kevin Dorst, Jeremy Goodman, Dan Greco, Brian Hedden, Dan Hoek, Sophie Horowitz, Jens Kipper, Harvey Lederman, Matt Mandelkern, Damien Rochford, Bernhard Salow, Miriam Shoenfield, Jonathan Vogel, and Ian Wells.

Epistemology has been a lively area of research at MIT, and over a wider range of time I have benefited from the almost constant flow of stimulating discussion, informal and in reading groups as well as seminars, with graduate students and colleagues. In addition to those already mentioned, I want to thank the following who have helped me to understand these issues, both during their time at MIT and later: Ray Briggs, Alex Byrne, Andy Egan, Adam Elga, Ned Hall, Caspar Hare, Justin Khoo, Sarah Moss, Milo Philips-Brown, Agustin Rayo, Ginger Schultheis, Jack Spencer, Jason Stanley, Eric Swanson, Zoltan Szabo, Roger White, Steve Yablo, and Seth Yalcin.

My debts to Timothy Williamson, Dorothy Edgington, and David Lewis will be evident throughout these papers. With each, there is the right mix of agreement and disagreement to make for fruitful discussion. Each has had a profound influence on my ideas.

Thanks, as always, to Peter Momtchiloff for his support and advice, and to David Balcarras for editorial help.

Details of First Publication

I thank the editors and publishers for permission to reprint the following previously published papers in this collection.

Chapter 1 is reprinted by permission from Springer Nature, *Philosophical Studies*, Volume 120, Issue 1, Robert Stalnaker, "On Logics of Knowledge and Belief," 169–99. Copyright © 2006.

Chapter 2 first appeared as Robert Stalnaker, "Luminosity and the KK Thesis" in *Externalism, Self-knowledge and Skepticism*, edited by Sanford C. Goldberg, 17–40, Cambridge University Press. Copyright © 2015. Reprinted by permission of Cambridge University Press.

Chapter 3 is reprinted by permission from Springer Nature, *Erkenntnis*, Volume 70, Issue 2, Robert Stalnaker, "Iterated Belief Revision," 189–209. Copyright © 2008.

Chapter 4 first appeared as Robert C. Stalnaker, "Modeling a Perspective on the World" in *About Oneself: De Se Thought and Communication*, edited by Manuel García-Carpintero and Stephan Torre, 121–37. Copyright © 2016. Reprinted by permission of Oxford University Press: https://global.oup.com/academic/product/about-oneself-9780198713265.

Chapter 7 also appears as Robert C. Stalnaker, "Expressivism and Propositions," forthcoming in *Unstructured Content*, edited by Dirk Kindermann, Andy Egan, and Peter van Elswyk, and is reprinted by permission of Oxford University Press.

Chapter 9 is reproduced with permission from Robert Stalnaker, "A Theory of Conditionals," in *Studies in Logical Theory*, edited by Nicholas Rescher, 98–112, Basil Blackwell. Copyright © 1968.

Chapter 10 first appeared as Robert C. Stalnaker, "Conditional Propositions and Conditional Assertions," in *Epistemic Modality* edited by Brian Weatherson and Andy Egan, 227–48. Copyright © 2011, and is reprinted by permission of Oxford University Press: https://global.oup.com/academic/product/epistemic-modality-9780199591589.

Chapter 11 also appears as Robert C. Stalnaker, "Counterfactuals and Probability," forthcoming in *Conditionals, Paradox and Probability: Themes from the Philosophy of Dorothy Edgington*, edited by Lee Walters and John Hawthorne, and is reprinted by permission of Oxford University Press.

Chapter 12 first appeared as Robert Stalnaker, "Counterfactuals and Humean Reduction," in *A Companion to David Lewis*, edited by Barry Loewer and Jonathan Schaffer, 411–24, Wiley-Blackwell. Copyright © 2015. Reprinted by permission of the editors.

The following chapters are published here for the first time:

"Reflection, Endorsement, Calibration"
"Rational Reflection and the Notorious Unmarked Clock"
"Contextualism and the Logic of Knowledge"
"Dispositions and Chance."

Introduction

More than thirty years ago I wrote a book called *Inquiry*. This was a great title for a philosophy book, with its allusion (or homage) to classic works in the empiricist tradition, and it was an appropriate title for the aspirations with which the book was written: its topic, I said in the preface, was the abstract structure of inquiry. But it is less clear that this was an appropriate title for what was actually accomplished in the book since it did not get much beyond preliminary setting up of the issues, and some exposition of and motivation for the formal apparatus that I planned to use to talk about the structure of inquiry. Before getting to the main issues, I had to explain and motivate my approach to the problem of intentionality, sketch and motivate the formal apparatus used to represent that approach (possible worlds semantics), and respond to problems that the approach faced. That took up most of the book. The rest of it focused mainly on another piece of apparatus needed to represent the dynamics of belief (a formal semantics for conditionals), and I was able to make only a start on a discussion of the role of this apparatus in forming and refining both rules for revising beliefs, and concepts for giving a theoretical description of the world.

I said at the time (again in the preface) that I had begun that project with a naïve hope that I could get to the bottom of the problems I was concerned with, but that I had learned that the bottom was further down than I thought and so was then prepared only to make a preliminary progress report. The present collection is a further progress report on the same project, but I have changed my mind about getting to the bottom of things. I've decided there is no bottom: the best we can do in philosophy is to chip away at bits and pieces of the problems. We can paint impressionistic big pictures that we hope will get one to see the issues in a new and better way, and we can construct models that achieve precision only at the cost of idealization and simplification, but that we hope will throw some light on the phenomena. That may be enough to count as progress.

One gains some perspective from putting a collection together, seeing connections and recurrent themes that one had not noticed when working on the individual papers. One thing that stood out for me as I selected papers for this collection, and added to them to fill in gaps, was the continuity with the earlier book, even though all but one of the papers in this collection were written more than thirty years after *Inquiry* was published. This collection also has two parts, papers on knowledge and papers on conditionals, and these papers discuss the same themes discussed in the two parts of the earlier book. The focus of the first part has changed from belief to knowledge, but I have come to see that the problem of intentionality (at least on my way of approaching it) is essentially the same as the problem of characterizing knowledge. Knowledge whether ϕ, according to a slogan I like, is the capacity to make one's actions depend on whether ϕ. Knowledge is a matter of causal sensitivity to facts that are the subject matter of one's knowledge. My earlier gestures at explaining intentionality took a similar form: I took belief and desire to be the basic intentional states, but argued that belief states get their intentional content from the information that they tend to be sensitive to (under certain normal conditions). Looking back from the later perspective of Timothy Williamson's general picture of epistemology, I came to appreciate that my account of intentionality is really a version of his "knowledge first" view: belief is what would be knowledge if the relevant normal conditions in fact obtained, or to put it the other way around, knowledge is full belief when it is non-defective.

The papers in the second half of this collection develop further the ideas about conditionals that are sketched in the last three chapters of *Inquiry*: their role in epistemology, the metaphysical status of the propositions they express, and their relation to probabilistic concepts, both credence and chance. In the earlier book I sketched and defended what I called the projection strategy for explaining objective modal concepts as a kind of projection of epistemic states and policies onto the world, arguing that this strategy helped to explain the relation between the two kinds of conditionals (indicative and subjunctive). The strategy has its roots in Hume, but I contrasted it with the kind of reductionist Humean project that David Lewis developed. On my anti-reductionist account, the result of the projection is concept formation that refines our descriptive resources for distinguishing between the possible ways that the world might be. In the papers in the second part of this collection I look in more detail at these same issues.

I will sketch in broad strokes the picture of epistemology that is guiding me, and then try to put the individual papers in context by saying how I see their relation to this big picture.

The main problem of epistemology is to explain how we cognitive beings are able to find our way about in the world: how do we acquire and use the information about our environment that we need to succeed in it? Even the simplest animals acquire and use information, and they (along with simple artifacts) provide useful models of knowledge, but one thing that distinguishes the kind of cognitive beings we are from these simple cases is that we can reflect on ourselves as cognitive beings; part of the information we are able to acquire and use is information about our own place in the world—information about how we are able to acquire and use information. The point is not just that one of the inquiries we can engage in is epistemology. It is that any inquiry will involve at least implicit consideration of the methods we are using to reach the conclusions we reach, and when we are surprised—when something we took ourselves to know is shown to be false—we are forced to reflect on what went wrong: what assumption we were making, perhaps implicitly, about our epistemic connections to the world, and what changes we need to make in those assumptions to recover from our mistakes. The upshot is that one of the important channels of information involved in our acquisition and revision of knowledge is information about ourselves and our place in the world.

A second distinctive feature of the kind of complex cognitive beings that we are is that we are social creatures who rely on the knowledge of others. That is, knowers other than ourselves are involved in the channels through which we receive information. Critical reflection on these channels of information will be reflection, from a third person perspective, on how it is that cognitive beings like ourselves are able to find their way around, what *their* sources of information are, and what the world is like from their perspectives.

So, we develop a conception of the world a prominent part of which is ourselves and others like us—rational agents who are developing and refining a conception of the world they are in. Clarifying a conception of this kind will involve considering different perspectives on the world, and relations between those perspectives. In a sense, we are looking at ourselves from the outside, as agents whose interactions with nature and with other agents are part of an objective world to be described and explained. But we also recognize that we aren't really outside: Our third-person view of ourselves is developed and refined within the world, from perspectives within it. Getting clear about the relationships between different cognitive perspectives—that of the theorist, that of oneself at the moment, that of oneself at remembered and anticipated times, and that of others—is one of the aims in many of the papers in this collection.

The picture is a naturalistic one that sees cognitive beings as part of the natural world. Taking a page from Hume, this naturalistic picture gives no role to pure reason, beyond the requirements of consistency and coherence, in its account of inductive knowledge. I take the upshot of Hume's skeptical argument that reason cannot justify inductive practice, and his judgment that all reasoning about matters of fact is based on cause and effect to be something like this: we can't separate the task of developing and justifying rules for finding out about the world from the substantive task of developing a view about what the world is like. We approach both tasks from within, criticizing and refining the methods and beliefs that we find ourselves with.

To develop and sharpen this picture, it helps to have some formal tools. The book begins, in chapter 1, with a review of a formal semantics, in the possible-worlds framework, of knowledge and belief. A feature of this way of modeling knowledge (pioneered by Jaakko Hintikka) is that it provides a way of representing propositions about what an agent knows as propositions that are themselves the contents of knowledge. In the early theories of this kind, just a single knower was modeled, but the framework naturally extends to a theory with multiple knowers who have knowledge and beliefs about the knowledge and beliefs of each other, so this is an appropriate framework for developing the general picture sketched above. In the particular version of a model theory for knowledge and belief that I sketch in this chapter, assumptions are made that permit belief to be reduced knowledge, which is appropriate to the "knowledge first" ideology that is implicit in the information-theoretic picture of knowledge. And it provides a framework for clarifying questions about further constraints on the relation between knowledge and (full) belief (where your full beliefs are, roughly, the propositions you rightly or wrongly take yourself to know). Belief, in this sense, and knowledge will coincide when one is right about everything, but we can consider how much we can generalize about what an agent knows when some of what she takes herself to know is false. That is, what can one say, at this level of abstraction, about the extent to which errors in some of our knowledge claims infect others of our knowledge claims, and the extent to which some of these knowledge claims can be isolated from others. I pose this question in the model theoretic framework, and draw some connections between answers to it and proposals that arose in the very different post-Gettier project of trying to analyze knowledge in terms of true belief, plus some further condition.

The logic of knowledge sketched in the first chapter makes some transparency or luminosity assumptions that are controversial. While, as I have suggested, my picture conforms in many ways with Williamson's externalist epistemology, it diverges sharply from his on the issue of luminosity. But I argue in the second chapter, with the help of some simple models of information-carrying devices, that these assumptions—particularly the assumption that one who knows that ϕ is in a position to know that she knows it—can be reconciled with a thoroughly externalist conception of knowledge.

A second piece of formal apparatus that is relevant to the dynamic dimension of the picture is a formal theory of belief revision. The main task of the standard belief revision theory is to specify constraints on the way a subject is disposed to change his overall belief state as a result of discovering that some prior full belief is false. The third chapter is a critical discussion of some attempts to extend the standard theory to give an account of the way one's belief revision policies, as well as one's beliefs, should change in response to the discovery that a prior belief is false. Some elegant theories of iterated belief revision have been proposed, and they help to clarify the terrain, but I argue that they all face counterexamples. Although the main points I make in this chapter are negative, the counterexamples point to the importance of meta-information—the agent's knowledge and beliefs about her own epistemic situation—in belief revision. A fully satisfactory belief revision will involve an explanation of why revision was required—of what deviation from normal condition led one to take oneself to know something that one did not know—but often one learns one was mistaken without learning why, and this complicates the process of rational belief revision.

The fourth chapter is about the way self-locating knowledge and belief should be represented. On the picture of cognitive beings that I am working with, all knowledge is, in a sense, self-locating since all of an agent's representations get their content from that agent's relation to the things those representations are about. I argue that the standard way of thinking about self-locating belief, which distinguishes sharply between knowing what possible world you are in and knowing where you are in the world, is confused. You can know what country you are in without knowing where you are in the country, but (I argue) ignorance of where you are, or what time it is, is always ignorance about what the world is like, which is to say, about what possible world you are in. Models that recognize this can help give a clearer view of the way we think about the relations between epistemic perspectives, since charting those relations requires calibrating the relations between the contents of the attitudes of different agents, and of the same agent at different times.

Full belief, on the picture I am developing, is what one takes oneself to know, but a cognitive state will also include degrees of partial belief, and a "knowledge first" epistemology must concern itself with these more fine-grained states as well. Questions about the relationships between different cognitive perspectives will include questions about the relationships between the credence functions of different agents, and of the same agent at different times. Chapters 5 and 6 are about reflection or deference principles: principles that state constraints on an agent's credences about a credence function other than her own at that time: about her own anticipated or remembered credences, about the credences of another agent, or about the credences that she ideally ought to have. I argue in chapter 5 that reflection principles about oneself at other times or about others can be defended on the condition that one *endorses* those other credence functions, which means that one judges that they are the right credences for the relevant agent to have. In chapter 6, I explore a puzzle about the attitudes that an agent should have about the rationality of her own present attitudes.

The conception that our cognitive agent is forming and refining has a normative dimension. Her inquiries ask what the world is like, while at the same time asking what rules and procedures she should adopt to form beliefs and partial beliefs about what the world is like. Chapter 7 focuses on the normative or practical dimension. It sketches a framework, developed by Allan Gibbard, for representing a mix of normative and factual beliefs. While I endorse Gibbard's expressivist framework, I reject his own interpretation of that framework, arguing that it blurs the line between a realist and an expressivist conception of norms. This chapter is mostly about norms in general, but in the last part I look at some ways in which this framework helps to clarify more specific questions about epistemic norms, and the ways their application is constrained by facts.

The information-theoretic conception of knowledge is necessarily a contextualist conception for the following reason: Knowledge claims can be made only against a background of factual presuppositions since, on that conception, knowledge is based on naturalistic causal relations between a knower and the environment that is known. But the presuppositions relative to which knowledge is defined can themselves always be questioned, and addressing those questions requires a shift in the context. Chapter 8 develops the information-theoretic version of contextualism about knowledge, comparing and contrasting it with a contextualist theory developed by David Lewis that has a very different motivation.

The second part of the book contains papers that focus on conditional propositions: their role in representing epistemic policies, their contribution to the theoretical resources for describing the world, and their connections with other objective modal notions such as dispositional properties and chance.

As noted above, all but one of the papers were written in the last ten or twelve years. The exception is chapter 9, my first paper on conditionals, which is now about fifty years old. It is included here since it is the starting point of a project that led to a progress report sixteen years later, and to another one now. The formal logic and semantics for conditionals developed in this paper were similar to those in a theory being developed independently at about the same time by David Lewis, but the philosophical ideas guiding our two theories were very different. My project was less ambitious than Lewis's, disclaiming any attempt to provide a reductive analysis. The aim was just to clarify the formal structure of a concept and to provide the semantic apparatus with some intuitive motivation. The theory's aim was to do for counterfactual conditionals what Kripke's possible worlds semantics did for the concepts of necessity and possibility, which was manifestly not a reduction of modal concepts to something else. The projection strategy was not explicit in this early paper, but it was prefigured in the appeal to Ramsey's explanation of indicative conditionals to motivate an analysis that had counterfactual conditionals as its main target. Ramsey's suggestion was about how to decide whether to accept a conditional: add the antecedent, hypothetically, to your stock of beliefs, and accept the conditional if and only if the resulting hypothesized stock of beliefs implies the consequent. My question (after extending Ramsey's suggestion to cover cases where the antecedent was incompatible with your beliefs) was this: What should the truth-conditions for a conditional proposition be if this is a good way of deciding whether to accept it? A selection function from a possible world plus a proposition to a possible world in which the proposition is true was thought of as an ontological analogue of a function from a state of belief plus a proposition to a hypothetical state of belief, and so as a kind of projection of a relation between cognitive states onto the world.

While my account of conditionals presupposed that the problem was to give truth-conditions for conditional propositions, others at about this time were arguing that one should explain conditional sentences as sentences for performing a distinctive kind of speech act, or for representing a distinctive kind of conditional attitude. Ernest Adams developed a probabilistic semantics that began with the idea that a conditional is assertable when the probability of the consequent, conditional on the antecedent, is high, and Dorothy Edgington, building on Adams's work, developed some powerful arguments for a non-propositional account. Some philosophers such as Allan Gibbard gave a divided account of conditionals, siding with Edgington in giving a conditional assertion account of indicative conditionals, but with the propositionalists on subjunctive conditionals. Both Edgington and I aimed for unified accounts of the two kinds of conditionals, but accounts that allowed for and explained the differences. In chapters 10 and 11 I defend an ecumenical approach to the dispute between propositionalists and those who want to explain conditionals in terms of conditional speech acts and attitudes, arguing in chapter 10 that the conditional assertion analysis can be formulated as a limiting case of the propositional analysis, and that it is useful to do so since it helps to chart the connections and continuities between indicative and subjunctive conditionals. Chapter 11 focuses on Edgington's account of counterfactuals, and on the relations between counterfactuals and objective probability. I argue in this chapter that the propositional account can allow for indeterminacy, and can better explain the phenomena she uses to defend her account of the role of counterfactuals in epistemic reasoning.

Chapter 12 develops and criticizes David Lewis's reductive analysis of counterfactuals, and the Humean supervenience metaphysics that underlies it. Lewis's Humean theory contrasts with the usual empiricist defense of a Humean metaphysics, which takes the supervenience base to be observational or phenomenal concepts. For Lewis, the properties to which all else is to be reduced are the fundamental properties of physics. His picture also contrasts with the picture I have been developing, which puts causal notions at the center both of the descriptive resources for describing the world, and of the rules for our practice of learning about the world. Lewis separates metaphysical questions concerning what there is a fact of the matter about from epistemological questions about the proper rules for learning about those facts, while I try to draw conceptual connections between the two.

The final chapter, about dispositions and chance, is the most detailed discussion of the projection strategy, and the most explicit development of an application of it, the application to the concept of objective chance. I argue in the conclusion of this chapter that while the general picture is based on constitutive conceptual connections between epistemic rules and descriptive theoretical concepts (as phenomenalist and verificationist theories were), it is nevertheless a thoroughly realist metaphysical picture.

PART I
Knowledge

1

On the Logics of Knowledge and Belief

1. Introduction

Formal epistemology, or at least the approach to formal epistemology that develops a logic and formal semantics of knowledge and belief in the possible worlds framework, began with Jaakko Hintikka's book *Knowledge and Belief* (Hintikka1962). Hintikka's project sparked some discussion of issues about iterated knowledge (does knowing imply knowing that one knows?) and about "knowing who," and quantifying into knowledge attributions. Much later, this kind of theory was taken up and applied by theoretical computer scientists and game theorists.[1] The formal semantic project gained new interest when it was seen that it could be applied to contexts with multiple knowers, and used to clarify the relation between epistemic and other modal concepts.

Edmund Gettier's classic refutation of the Justified True Belief analysis of knowledge (Gettier 1963) was published at about the same time as Hintikka's book, and it immediately spawned an epistemological industry—a project of attempting to revise the refuted analysis by adding further conditions to meet the counterexamples. Revised analyses were met with further counterexamples, followed by further refinements. This kind of project flourished for some years, but eventually became an internally driven game that was thought to have lost contact with the fundamental epistemological questions that originally motivated it. This way of approaching epistemological questions now seems hopelessly out of date, but I think there may still be some insights to be gained by looking back, if not at the details of the analyses, at some of the general strategies of analysis that were deployed.

There was little contact between these two very different epistemological projects. The first had little to say about substantive questions about the relation between knowledge, belief, and justification or epistemic entitlement, or about traditional epistemological issues, such as skepticism. The second project ignored questions about the abstract structure of epistemic and doxastic states. But I think some of the abstract questions about the logic of knowledge connect with traditional questions in epistemology, and with the issues that motivated the attempt to find a definition of knowledge. The formal semantic framework provides the resources to construct models that may help to clarify the abstract relationship between the concept of

[1] See Fagin et al. 1995 and Battigalli & Bonanno 1999 for excellent surveys of the application of logics of knowledge and belief in theoretical computer science and game theory.

knowledge and some of the other concepts (belief and belief revision, causation and counterfactuals) that were involved in the post-Gettier project of defining knowledge. And some of the examples that were originally used in the post-Gettier literature to refute a proposed analysis can be used in a different way in the context of formal semantic theories: to bring out contrasting features of some alternative conceptions of knowledge, conceptions that may not provide plausible analyses of knowledge generally, but that may provide interesting models of knowledge that are appropriate for particular applications, and that may illuminate, in an idealized way, one or another of the dimensions of the complex epistemological terrain.

My aim in this chapter will be to bring out some of the connections between issues that arise in the development and application of formal semantics for knowledge and belief and more traditional substantive issues in epistemology. The chapter will be programmatic, pointing to some highly idealized theoretical models, some alternative assumptions that might be made about the logic and semantics of knowledge, and some of the ways in which they might connect with traditional issues in epistemology, and with applications of the concept of knowledge. I will bring together and review some old results, and make some suggestions about possible future developments. After a brief sketch of Hintikka's basic logic of knowledge, I will discuss, in section 2, the S5 epistemic models that were developed and applied by theoretical computer scientists and game theorists, models that, I will argue, conflate knowledge and belief. In section 3, I will discuss a basic theory that distinguishes knowledge from belief and that remains relatively noncommittal about substantive questions about knowledge, but that provides a definition of belief in terms of knowledge. This theory validates a logic of knowledge, S4.2, that is stronger than S4, but weaker than S5. In the remaining four sections, I will consider some alternative ways of adding constraints on the relation between knowledge and belief that go beyond the basic theory: in section 4 I will consider the S5 partition models as a special case of the basic theory; in section 5 I will discuss the upper and lower bounds to an extension of the semantics of belief to a semantics for knowledge; in section 6 I will discuss a version of the defeasibility analysis of knowledge, and in section 7 a simplified version of a causal theory.

The basic idea that Hintikka developed, and that has since become familiar, was to treat knowledge as a model operator with a semantics that parallels the possible worlds semantics for necessity. Just as necessity is truth in all possible worlds, so knowledge is truth in all *epistemically* possible worlds. The assumption is that to have knowledge is to have a capacity to locate the actual world in logical space, to exclude certain possibilities from the candidates for actuality. The epistemic possibilities are those that remain after the exclusion, those that the knower cannot distinguish from actuality. To represent knowledge in this way is of course not to provide any kind of reductive analysis of knowledge, since the abstract theory gives no substantive account of the criteria for determining epistemic possibility. The epistemic possibilities are defined by a binary accessibility relation between possible worlds that is a primitive component of an epistemic model. (Where x and y are possible worlds, and "R" is the accessibility relation, "xRy" says that y is epistemically possible for the agent in world x.) The idea was to give a precise representation of the structure of an epistemic state that was more or less neutral about more substantive questions about what constitutes knowledge, but that sharpened questions about the logic of knowledge. This form of representation was, however, far from innocent, since it required, from the start, an extreme idealization: Even in its most neutral form, the framework required the assumption that knowers know all logical truths and all of the consequences of their knowledge, since no matter how the epistemically possible worlds are selected, all logical truths will be true in all of them, and for any set of propositions true in all of them, all of their logical consequences will also be true in all of them. There are different ways of understanding the character of this idealization: on the one hand, one might say that the concept of knowledge that is being modeled is knowledge in the ordinary sense, but that the theory is intended to apply only to idealized knowers—those with superhuman logical capacities. Alternatively, one might say that the theory is intended to model an idealized sense of knowledge—the information that is implicit in one's knowledge—that literally applies to ordinary knowers. However the idealization is explained, there remain the questions whether it is fruitful to develop a theory that requires this kind of deviation from reality, and if so why.[2] But I think these questions are best answered by looking at the details of the way such theories have been, and can be developed.

The most basic task in developing a semantics for knowledge in the possible worlds framework is to decide on the properties of the epistemic accessibility relation. It is clear that the relation should be reflexive, which is necessary to validate the principle that knowledge implies truth, an assumption that is just about the only principle of a logic of knowledge that is uncontroversial. Hintikka argued that we should also assume that the relation is transitive, validating the much more controversial principle that knowing implies knowing that one knows. Knowing and knowing that one knows are, Hintikka claimed, "virtually equivalent." Hintikka's reasons for this conclusion were not completely clear. He did not want to base it on a capacity for introspection: he emphasized that his reasons were logical rather than psychological. His proof of the KK principle rests on the following principle: If $\{K\phi, {\sim}K{\sim}\psi\}$ is consistent, then $\{K\phi, \psi\}$ is consistent, and it is clear that if one grants this principle, the KK principle immediately follows.[3] The reason for accepting this principle seems to be something like this: Knowledge requires conclusive reasons for belief, reasons that would not be defeated by any information compatible with what is known. So, if one knows that ϕ while ψ is compatible with what one knows, then the truth of ψ could not defeat one's claim to know that ϕ. This argument, and other considerations for and against the KK principle deserve more careful scrutiny. There is a tangle of important and interesting issues underlying the question whether one should accept the KK principle and the corresponding semantics, and some challenging arguments that need to be answered if one does.[4] I think the principle can be

[2] I explored the problem of logical omniscience in two papers, Stalnaker 1991 and 1999b. I don't attempt to solve the problem in either paper, but only to clarify it, and to argue that it is a genuine problem, and not an artifact of a particular theoretical framework.

[3] Substituting "$\sim K\phi$" for ψ, and eliminating a double negation, the principle says that if $\{K\phi, {\sim}KK\phi\}$ is consistent, then $\{K\phi, {\sim}K\phi\}$ is consistent.

[4] See especially, Williamson 2000 for some reasons to reject the KK principle. I respond to Williamson's main argument in Stalnaker 2015, reprinted as chapter 2 of this book.

defended (in the context of the idealizations we are making), but I will not address this issue here, provisionally following Hintikka in accepting the KK principle, and a semantics that validates it.

The S4 principles (Knowledge implies truth, and knowing implies knowing that one knows) were as far as Hintikka was willing to go. He unequivocally rejects the characteristic S5 principle that if one lacks knowledge, then one knows that one lacks it ("unless you happen to be as sagacious as Socrates"[5]), and here his reasons seem to be clear and decisive:

> The consequences of this principle, however, are obviously wrong. By its means (together with certain intuitively acceptable principles) we could, for example, show that the following sentence is self sustaining:
>
> $$p \rightarrow K_a P_a P.^{6} \qquad (13)$$

The reason that (13) is clearly unacceptable, as Hintikka goes on to say, is that it implies that one could come to know by reflection alone, of any truth, that it was compatible with one's knowledge. But it seems that a consistent knower might believe, and be justified in believing, that she knew something that was in fact false. That is, it might be, for some proposition ϕ, that ~ϕ, and BKϕ. In such a case, if the subject's beliefs are consistent, then she does not believe, and so does not know, that ~ϕ is compatible with her knowledge. That is, ~K~Kϕ, along with ~ϕ, will be true, falsifying (13).

2. Partition Models

Despite Hintikka's apparently decisive argument against the S5 principle, later theorists applying epistemic logic and semantics, both in theories of distributive computer systems and in game theory assumed that S5 was the right logic for (an idealized concept of) knowledge, and they developed semantic models that seem to support that decision. But while such models, properly interpreted, have their place, I will argue that the theorists defending them conflated knowledge and belief in a way that has led to some conceptual confusion, and that they have abstracted away from some interesting problems within their intended domains of application that more general models might help to clarify. But before getting to this issue, let me first take note of another way that more recent theorists have modified, or generalized, Hintikka's original theory.

Hintikka's early models were models of the knowledge of a single knower, but much of the later interest in formal epistemic models derives from a concern with situations in which there are multiple knowers who may know or be ignorant about the knowledge and ignorance of the others. While Hintikka's early work did not give explicit attention to the interaction of different knowers, the potential to do so is implicit in his theory. Both the logic and the semantics of the knowledge of a single

[5] Hintikka 1962, 106.

[6] Ibid., 54. In Hintikka's notation, "P_a" is the dual of the knowledge operator, "K_a": "$\sim K_a \sim$". I will use "M" for ~K~).

knower generalize in a straightforward way to a model for multiple knowers. One needs only a separate knowledge operator for each knower, and in the semantics, a separate relation of epistemic accessibility for each knower that interprets the operator. One can also introduce, for any group of knowers, an operator for the *common* knowledge shared by the member of the group, where a group has common knowledge that ϕ if and only if all know that ϕ, all know that all know that ϕ, all know that all know that all know, etc. all the way up. The semantics for the common knowledge operator is interpreted in terms of an accessibility relation that is definable in terms of the accessibility relations for the individual knowers: the common-knowledge accessibility relation for a group G is the transitive closure of the set of epistemic accessibility relations for the members of that group.[7] If R^G is this relation, then the knowers who are members of G have common knowledge that ϕ (in possible world x) iff ϕ is true in all possible worlds that are R^G related to world x. The generalization to multiple knowers and to common knowledge works the same way, whatever assumptions one makes about the accessibility relation, and one can define notions of common belief in an exactly analogous way. The properties of the accessibility relations for common knowledge and common belief will derive from the properties of the individual accessibility relations, but they won't necessarily be the same as the properties of the individual accessibility relations. (Though if the logic of knowledge is S4 or S5, then the logic of common knowledge will also be S4 or S5, respectively).

Theoretical computer scientists have used the logic and semantics for knowledge to give abstract descriptions of distributed computer systems (such as office networks or email systems) that represent the distribution and flow of information among the components of the system. For the purpose of understanding how such systems work and how to design protocols that permit them to accomplish the purposes for which they were designed, it is useful to think of them as communities of interacting rational agents who use what information they have about the system as a whole to serve their own interests, or to play their part in a joint project. And it is useful in turn for those interested in understanding the epistemic states of rational agents to think of them as analogues of the kind of simplified models that theoretical computer scientists have constructed.

A distributed system consists of a set of interconnected components, each capable of being in a range of local states. The way the components are connected, and the rules by which the whole system works, constrain the configurations of states of the individual components that are possible. One might specify such a system by positing a set of n components and possible local states for each. One might also include a component labeled "nature" whose local states represent information from outside the system proper. *Global* states will be n-tuples of local states, one for each component, and the model will also specify the set of global states that are *admissible.* Admissible global states are those that are compatible with the rules governing the

[7] More precisely, if R^i is the accessibility relation for knower i, then the common-knowledge accessibility relation for a group G is defined as follows: xR^Gy iff there is a sequence of worlds, $z_1, \ldots z_n$ such that $z_1 = x$ and $z_n = y$ and for all j between 1 and $n-1$, there is a knower $i \in G$, such that $z_jR^i\, z_{j+1}$.

way the components of the system interact. The admissible global states are the possible worlds of the model. This kind of specification will determine, for each local state that any component might be in, a set of global states (possible worlds) that are compatible with the component being in that local state. This set will be the set of epistemically possible worlds that determines what the component in that state knows about the system as a whole.[8] Specifically, if "*a*" and "*b*" denote admissible global states, and "a_i" and "b_i" denote the ith elements of *a* and *b*, respectively (the local states of component *i*), then global world-state *b* is epistemically accessible (for *i*) to global world-state *b* if and only if $a_i = b_i$. So, applying the standard semantic rule for the knowledge operator, component (or knower) *i* will know that ϕ, in possible world *a*, if and only if ϕ is true in all possible worlds in which *i* has the same local state that it has in world-state *a*. One knows that ϕ if one's local state carries the information that ϕ.[9]

Now it is obvious that this epistemic accessibility relation is an equivalence relation, and so the logic for knowledge in a model of this kind is S5. Each of the epistemic accessibility relations partitions the space of possible worlds, and the cross-cutting partitions give rise to a simple and elegant model of common knowledge, also with an S5 logic. Game theorists independently developed this kind of partition model of knowledge and have used such models to bring out the consequences of assumptions about common knowledge. For example, it can be shown that in certain games, players will always make certain strategy choices when they have common knowledge that all players are rational. But as we have seen, Hintikka gave reasons for rejecting the S5 logic for knowledge, and the reasons seemed to be decisive. It is clear that a consistent and epistemically responsible agent might take herself to know that ϕ in a situation in which ϕ was in fact false. Because knowledge implies truth, it would be false, in such a case, that the agent knew that ϕ, but the agent could not know that she did not know that ϕ without having inconsistent beliefs. If such a case is possible, then there will be counterexamples to the S5 principle ($\sim K\phi \rightarrow K\sim K\phi$). That is, the S5 principles require that rational agents be immune to error. It is hard to see how any theory that abstracts away from the possibility of error could be relevant to epistemology, an enterprise that begins with skeptical arguments using scenarios in which agents are systematically mistaken and that seeks to explain the relation between knowledge and belief, presupposing that these notions do not coincide.

[8] A more complex kind of model would specify a set of admissible *initial* global states, and a set of transition rules taking global states to global states. The possible worlds in this kind of model are the admissible global *histories*—the possible ways that the system might evolve. In this kind of model, one can represent the distribution of information, not only about the current state of the system, but also about how it evolved, and where it is going. In the more general model, knowledge states are time-dependent, and the components may have or lack information, not only about which possible world is actual, but also about where (temporally) it is in a given world. The dynamic dimension, and the parallels with issues about indexical knowledge and belief, are part of the interest of the distributed systems models, but I will ignore these issues here.

[9] Possible worlds, on this way of formulating the theory, are not primitive points, as they are in the usual abstract semantics, but complex objects—sequences of local states. But an equivalent formulation might begin with a given set of primitive (global) states, together with a set of equivalence relations, one for each knower, and one for "nature." The local states could then be defined as the equivalence classes.

Different theorists have different purposes, and it is not immediately obvious that the models of knowledge that are appropriate to the concerns of theoretical computer scientists and game theorists need be relevant to issues in epistemology. But I think that the possibility of error, and the differences between knowledge and belief are relevant to the intended domains of application of those models, and that some of the puzzles and problems that characterize epistemology are reflected in problems that may arise in applying those theories.

As we all know too well, computer systems sometimes break down or fail to behave as they were designed to behave. In such cases, the components of a distributed system will be subject to something analogous to error and illusion. Just as the epistemologist wants to explain how and when an agent knows some things even when he is in error about others, and is interested in methods of detecting and avoiding error, so the theoretical computer scientist is interested in the way that the components of a system can avoid and detect faults, and can continue to function appropriately even when conditions are not completely normal. To clarify such problems, it is useful to distinguish knowledge from something like belief.

The game theorist, or any theorist concerned with rational action, has a special reason to take account of the possibility of false belief, even under the idealizing assumption that in the actual course of events, everyone's beliefs are correct. The reason is that decision theorists and game theorists need to be concerned with causal or counterfactual possibilities, and to distinguish them from epistemic possibilities. When I deliberate, or when I reason about why it is rational to do what I know that I am going to do, I need to consider possible situations in which I make alternative choices. I know, for example, that it would be irrational to cooperate in a one-shot prisoners' dilemma because I know that in the counterfactual situation in which I cooperate, my payoff is less than it would be if I defected. And while I have the capacity to influence my payoff (negatively) by making this alternative choice, I could not, by making this choice, influence your prior beliefs about what I will do; that is, your prior beliefs will be the same, in the counterfactual situation in which I make the alternative choice, as they are in the actual situation. Since you take yourself (correctly, in the actual situation) to know that I am rational, and so that I will not cooperate, you therefore also take yourself to know, in the counterfactual situation I am considering, that I am rational, and so will not cooperate. But in that counterfactual situation, you are wrong—you have a false belief that you take to be knowledge. There has been a certain amount of confusion in the literature about the relation between counterfactual and epistemic possibilities, and this confusion is fed, in part, by a failure to make room in the theory for false belief.[10]

Even in a context in which one abstracts away from error, it is important to be clear about the nature of the idealization, and there are different ways of understanding it that are sometimes confused. But before considering the alternative ways of making the S5 idealization, let me develop the contrast between knowledge and belief, and the relation between them, in a more general setting.

[10] These issues are discussed in Stalnaker 1996.

3. Belief and Knowledge

Set aside the S5 partition models for the moment, and consider, from a more neutral perspective, the logical properties of belief, and the relation between belief and knowledge. It seems reasonable to assume, at least in the kind of idealized context we are in, that agents have introspective access to their beliefs: if they believe that ϕ, then they know that they do, and if they do not, then they know that they do not. (The S5, "negative introspection" principle, $(\sim K\phi \rightarrow K\sim K\phi)$, was problematic for knowledge because it is in tension with the fact that knowledge implies truth, but the corresponding principle for belief does not face this problem.) It also seems reasonable to assume that knowledge implies belief. Given the fact that our idealized believers are logically omniscient, we can assume, in addition, that their beliefs will be consistent. Finally, to capture the fact that our intended concept of belief is a strong one—subjective certainty—we assume that believing implies believing that one knows. So, our logic of knowledge and belief should include the following principles in addition to those of the logic S4:

(PI)	$\vdash B\phi \rightarrow KB\phi$	positive introspection
(NI)	$\vdash \sim B\phi \rightarrow K\sim B\phi$	negative introspection
(KB)	$\vdash K\phi \rightarrow B\phi$	knowledge implies belief
(CB)	$\vdash B\phi \rightarrow \sim B\sim\phi$	consistency of belief
(SB)	$\vdash B\phi \rightarrow BK\phi$	strong belief

The resulting combined logic for knowledge and belief yields a pure belief logic, KD45, which is validated by a doxastic accessibility relation that is serial, transitive, and Euclidean.[11] More interestingly, one can prove the following equivalence theorem: $\vdash B\phi \leftrightarrow MK\phi$ (using "M" as the epistemic possibility operator, "$\sim K\sim$"). This equivalence permits a more economical formulation of the combined belief-knowledge logic in which the belief operator is defined in terms of the knowledge operator. If we substitute "MK" for "B" in our principle (CB), we get $MK\phi \rightarrow KM\phi$, which, if added to S4 yields the logic of knowledge, S4.2. All of the other principles listed above (with "MK" substituted for "B") are theorems of S4.2, so this logic of knowledge by itself yields a combined logic of knowledge and belief with the appropriate properties.[12]

The assumptions that are sufficient to show the equivalence of belief with the epistemic possibility of knowledge (one believes that ϕ, in the strong sense, if and only if it is compatible with one's knowledge that one knows that ϕ) might also be made for a concept of *justified* belief, although the corresponding assumptions will be more controversial. Suppose (1) one assumes that justified belief is a necessary

[11] KD45 adds to the basic modal system K the axioms (D), which is our (CB), (4) $B\phi \rightarrow BB\phi$, which follows immediately from our (PI) and (KB), and (5) $\sim B\phi \rightarrow B\sim B\phi$, which follows immediately from (NI) and (KB). The necessitation rule for B (If $\vdash\phi$, then $\vdash B\phi$) and the distribution principle $(B(\phi \rightarrow \psi) \rightarrow (B\phi \rightarrow B\psi))$ can both be derived from our principles.

[12] The definability of belief in terms of knowledge, and the point that the assumptions about the relation between knowledge and belief imply that the logic of knowledge should be S4.2, rather than S4, were first shown by Wolfgang Lenzen. See his classic monograph, Lenzen 1978.

condition for knowledge, and (2) one adopts an *internalist* conception of justification that supports the positive and negative introspection conditions (if one has justified belief that ϕ, one knows that one does, and if one does not, one knows that one does not), and (3) one assumes that since the relevant concept of belief is a strong one, one is justified in believing that ϕ if and only if one is justified in believing that one knows that ϕ. Given these assumptions, justified belief will also coincide with the epistemic possibility that one knows, and so belief and justified belief will coincide. The upshot is that for an internalist, a divergence between belief (in the strong sense) and justified belief would be a kind of internal inconsistency. If one is not fully justified in believing ϕ, one knows this, and so one knows that a necessary condition for knowledge that ϕ is lacking. But if one believes that ϕ, in the strong sense, then one believes that one knows it. So, one both knows that one lacks knowledge that ϕ, and believes that one has knowledge that ϕ.

The usual constraint on the accessibility relation that validates S4.2 is the following convergence principle (added to the transitivity and reflexivity conditions): if $x\mathrm{R}y$ and $x\mathrm{R}z$, then there is a w such that $y\mathrm{R}w$ and $z\mathrm{R}w$. But S4.2 is also sound and complete relative to the following stronger convergence principle: for all x, there is a y such that for all z, if $x\mathrm{R}z$, then $z\mathrm{R}y$. The weak convergence principle (added to reflexivity and transitivity) implies that for any *finite* set of worlds accessible to x, there is a single world accessible with respect to all of them. The strong convergence principle implies that there is a world that is accessible to *all* worlds that are accessible to x. The semantics for our logic of knowledge requires the stronger convergence principle.[13]

Just as, within the logic, one can define belief in terms of knowledge, so within the semantics, one can define a doxastic accessibility relation for the derived belief operator in terms of the epistemic accessibility relation. If "R" denotes the epistemic accessibility relation and "D" denotes the doxastic relation, then the definition is as follows: $x\mathrm{D}y =_{\mathrm{df}} (z)(x\mathrm{R}z \rightarrow z\mathrm{R}y)$. Assuming that R is transitive, reflexive, and strongly convergent, it can be shown that D will be serial, transitive, and Euclidean—the constraints on the accessibility relation that characterize the logic KD45.

One can also define, in terms of D, and so in terms of R, a third binary relation on possible worlds that is relevant to describing the epistemic situation of our ideal knower: Say that two possible worlds x and y are *epistemically indistinguishable* to an agent ($x\mathrm{E}y$) if and only if she has exactly the same beliefs in world x as she has in world y. That is, $x\mathrm{E}y =_{\mathrm{df}} (z)(x\mathrm{D}z \leftrightarrow y\mathrm{D}z)$. E is obviously an equivalence relation, and so any modal operator interpreted in the usual way in terms of E would be an S5 operator. But while this relation is definable in the semantics in terms of the epistemic accessibility relation, we cannot define, in the object language with just the knowledge operator, a modal operator whose semantics is given by this accessibility relation.

[13] The difference between strong and weak convergence does not affect the *propositional* modal logic, but it will make a difference to the quantified modal logic. The following is an example of a sentence that is valid in models satisfying strong convergence (along with transitivity and reflexivity) but not valid in all models satisfying weak convergence: $\mathrm{MK}((x)(\mathrm{MK}\phi \rightarrow \phi))$.

So the picture that our semantic theory paints is something like this: For any given knower i and possible world x, there is, first, a set of possible worlds that are subjectively indistinguishable from x, to i (those worlds that are E-related to x); second, there is a subset of that set that includes just the possible worlds compatible with what i *knows* in x (those worlds that are R-related to x); third, there is a subset of that set that includes just the possible worlds that are compatible with what i *believes* in x (those worlds that are D-related to x). The world x itself will necessarily be a member of the outer set and of the R-subset, but will not necessarily be a member of the inner D-subset. But if x is itself a member of the inner D-set (if world x is itself compatible with what i believes in x), then the D-set will coincide with the R-set.

Here is one way of seeing this more general theory as a generalization of the distributive systems models, in which possible world-states are sequences of local states: one might allow *all* sequences of local states (one for each agent) to count as possible world-states, but specify, for each agent, a subset of them that are *normal*—the set in which the way that agent interacts with the system as a whole conforms to the constraints that the system conforms to when it is functioning as it is supposed to function. In such models, two worlds, x and y, will be subjectively indistinguishable, for agent i (xE_iy), whenever $x_i = y_i$ (so the relation that was the epistemic accessibility relation in the unreconstructed S5 distributed systems model is the subjective indistinguishability relation in the more general models). Two worlds are related by the doxastic accessibility relation (xD_iy) if and only if $x_i = y_i$, *and in addition, y is a normal world, with respect to agent i.*[14] This will impose the right structure on the D and E relations, and while it imposes some constraints on the epistemic accessibility relation, it leaves it underdetermined. We might ask whether R can be defined in a plausible way in terms of the components of the model we have specified, or whether one might add some independently motivated components to the definition of a model that would permit an appropriate definition of R. This question is a kind of analogue of the question asked in the more traditional epistemological enterprise—the project of giving a definition of knowledge in terms of belief, truth, justification, and whatever other normative and causal concepts might be thought to be relevant. Transposed into the model theoretic framework, the traditional problem of adding to true belief further conditions that together are necessary and sufficient for knowledge is the problem of extending the doxastic accessibility relation to a reflexive relation that is the right relation (at least in the idealized context) for the interpretation of a knowledge operator. In the remainder of this chapter, I will consider several ways that this might be done, and look at the logics of knowledge that they validate.

[14] We observed in note 8 that an equivalent formulation of the S5 distributed systems models would take the global world-states as primitive, specifying an equivalence relation for each agent, and defining local states as equivalence classes of global states. In an equivalent formulation of this kind of the more general theory, the assumption that every sequence of local states is a possible world will be expressed by a recombination condition: that for every sequence of equivalence classes (one for each agent) there is a possible world that is a member of their intersection. I have suggested that a recombination condition of this kind should be imposed on game theoretic models (where the equivalence classes are types, represented by probability functions), defending it as a representation of the conceptual independence of the belief states of different agents.

4. Partition Models and the Basic Theory

One extreme way of defining the epistemic accessibility relation in terms of the resources of our models is to identify it with the relation of subjective indistinguishability, and this is one way that the S5 partition models have implicitly been interpreted. If one simply assumes that the epistemic accessibility relation is an equivalence relation, this will suffice for a collapse of our three relations into one. Subjective indistinguishability, knowledge, and belief will all coincide. This move imposes a substantive condition on knowledge, and so on belief, when it is understood in the strong sense as belief that one knows, a condition that is appropriate for the skeptic who thinks that we are in a position to have genuine knowledge only about our own internal states—states about which we cannot coherently be mistaken. On this conception of knowledge, one can have a false belief (in the strong sense) only if one is internally inconsistent, and so this conception implies a bullet-biting response to the kind of argument that Hintikka gave against the S5 logic for knowledge. Hintikka's argument was roughly this: S5 validates the principle that any proposition that is in fact false, is known by any agent to be compatible with his knowledge, and this is obviously wrong: The response suggested by the conception of knowledge that identifies knowledge with subjective indistinguishability is that if we assume that all we can know is how things seem to us, and also assume that we are infallible judges of the way things seem to us, then it will be reasonable to conclude that we are in a position to know, of anything that is in fact false, that we do not know it.

There is a less radical way to reconcile our basic theory of knowledge and belief with the S5 logic and the partition models. Rather than making more restrictive assumptions about the concept of knowledge, or about the basic structure of the model, one may simply restrict the intended domain of application of the theory to cases in which the agent in question has, in fact, only true beliefs. On this way of understanding the S5 models, the model theory does not further restrict the relations between the three accessibility relations, but instead assumes that the *actual* world of the model is a member of the inner D-set.[15] This move does not provide us with a way to define the epistemic accessibility relation in terms of the other resources of the

[15] In most formulations of a possible-worlds semantics for propositional modal logic, a *frame* consists simply of a set of worlds and an accessibility relation. A model on a frame determines the truth values of sentences, relative to each possible world. On this conception of a model, one cannot talk of the truth of a sentence in a model, but only of truth *at a world* in a model. Sentence validity is defined, in formulations of this kind, as truth in all worlds in all models. But in some formulations, including in Kripke's original formal work, a frame (or model structure, as Kripke called it at the time) included, in addition to a set of possible worlds and an accessibility relation, a designated possible world—the actual world of the model. A sentence is true in a model if it is true in the designated actual world, and valid if true in all models. This difference in formulation was a minor detail in semantic theories for most of the normal modal logics, since any possible world of a model might be the designated actual world without changing anything else. So, the two ways of defining sentence validity will coincide. But the finer-grained definition of a frame allows for theories in which the constraints on R, and the semantic rules for operators, make reference to the actual world of the model. In such theories, truth in all worlds in all models may diverge from truth in all models, allowing for semantic models of logics that fail to validate the rule of necessitation.

model; but what it does is to stipulate that the actual world of the model is one for which the epistemic accessibility relation is determined by the other components. (That is, the set of worlds *y* that are epistemically accessible to the actual world is determined.) Since the assumptions of the general theory imply that all worlds outside the D-sets are epistemically inaccessible to worlds within the D-sets, and that all worlds within a given D-set are epistemically accessible to each other, the assumption that the actual world of the model is in a D-set will determine the R-set for the actual world, and will validate the logic S5.

So long as the object language that is being interpreted contains just one modal operator, an operator representing the knowledge of a single agent, the underdetermination of epistemic accessibility will not be reflected in the truth-values in a model of any expressible proposition. Since all possible worlds outside of any D-set will be invisible to worlds within it, one could drop them from the model (taking the set of all possible worlds to be those R-related to the actual world) without affecting the truth-values (at the actual world) of any sentence. This generated submodel will be a simple S5 model, with a universal accessibility relation. But as soon as one enriches the language with other modal and epistemic operators, the situation changes. In the theory with two or more agents, even if one assumes that all agents have only true beliefs, the full S5 logic will not be preserved. The idealizing assumption will imply that Alice's beliefs coincide with her knowledge (in the actual world), and that Bob's do as well, but it will not follow that Bob knows (in the actual world) that Alice's beliefs coincide with her knowledge. To validate the full S5 logic, in the multiple agent theory, we need to assume that it is not just true, but common knowledge that everyone has only true beliefs. This stronger idealization is needed to reconcile the partition models, used in both game theory and in distributed systems theory, with the general theory that allows for a distinction between knowledge and belief. But even in a context in which one makes the strong assumption that it is common knowledge that no one is in error about anything, the possible divergence of knowledge and belief, and the failure of the S5 principles to be *necessarily* true will show itself when the language of knowledge and common knowledge is enriched with non-epistemic modal operators, or in semantic models that represent the interaction of epistemic and non-epistemic concepts. In game theory, for example, an adequate model of the playing of a game must represent, not just the epistemic possibilities for each of the players, but also the capacities of players to make each of the choices that are open to that player, even when it is known that the player will not make some of those choices. One might assume that it is common knowledge that Alice will act rationally in a certain game, and it might be that it is known that Alice would be acting irrationally if she chose option X. Nevertheless, it would distort the representation of the game to deny that Alice has the option of choosing action X, and the counterfactual possibility in which she exercises that option may play a role in the deliberations of both Alice and the other players, whose knowledge that Alice will not choose option X is based on their knowledge of what she knows would happen if she did. So even if one makes the idealizing assumption that all agents have only true beliefs, or that it is common belief that everyone's beliefs are true, one should recognize the more general structure that distinguishes belief from knowledge, and that distinguishes both of these concepts from subjective indistinguishability. In the more general structure that recognizes these distinctions, the epistemic accessibility relation is underdetermined by the other relations.

5. Minimal and Maximal Extensions

So, our task is to say more about how to extend the relation D of doxastic accessibility to a relation R of epistemic accessibility. We know, from the assumption that knowledge implies belief, that in any model meeting our basic conditions on the relation between knowledge and belief, R will be an extension of D (for all x and y, if xDy, then xRy), and we know from the assumption that knowledge implies truth that the extension will be to a reflexive relation. We know by the assumption that belief is strong belief (belief that one knows) that R coincides with D, within the D-set (for all x and y, if xDx, then xRy if and only if xDy). What remains to be said is what determines, for a possible world x that is *outside* of a D-set, which other possible worlds outside that D-set are epistemically accessible to x. If some of my beliefs about what I know are false, what can be said about other propositions that I think that I know?

The assumptions of the neutral theory put clear upper and lower bounds on the answer to this question, and two ways to specify R in terms of the other resources of the model are to make the minimal or maximal extensions. The *minimal* extension of D would be the reflexive closure of D. On this account, the set of epistemically possible worlds for a knower in world x will be the set of doxastically accessible worlds, plus x. To make this minimal extension is to adopt the true belief analysis of knowledge, or in case one is making the internalist assumptions about justified belief, it would be to adopt the justified true belief analysis. The logic of true belief, S4.4, is stronger than S4.2, but weaker than S5.[16] The true belief analysis has its defenders, but most will want to impose stronger conditions on knowledge, which in our setting means that we need to go beyond the minimal extension of R.

It follows from the positive and negative introspection conditions for belief that for any possible world x, all worlds epistemically accessible to x will be subjectively indistinguishable from x (for all x and y, if xRy, then xEy) and this sets the upper bound on the extension of D to R. To identify R with the *maximal* admissible extension is to define it as follows: $x\text{R}y =_{\text{df}}$ either (xDx and xDy) or (not xDx and xEy). This account of knowledge allows one to know things that go beyond one's internal states only when *all* of one's beliefs are correct. The logic of this concept of knowledge, S4F, is stronger than S4.2, but weaker than the logic of the minimal extension, S4.4. The maximal extension would not provide a plausible account of knowledge in general, but it might be the appropriate idealization for a certain limited context. Suppose one's information all comes from a single source (an oracle), who you presume, justifiably, to be reliable. Assuming that all of its pronouncements are true, they give you knowledge, but in possible worlds in which any one of its pronouncements is false, it is an unreliable oracle, and so nothing it says

[16] See the appendix for a summary of all the logics of knowledge discussed, their semantics, and the relationships between them.

should be trusted. This logic, S4F, has been used as the underlying logic of knowledge in some theoretical accounts of a nonmonotonic logic. Those accounts don't provide an intuitive motivation for using this logic, but I think a dynamic model, with changes in knowledge induced by a single oracle who is presumed to be reliable, can provide a framework that makes intuitive sense of these nonmonotonic theories.[17]

6. Belief Revision and the Defeasibility Analysis

Any attempt to give an account of the accessibility relation for knowledge that falls between the minimal and maximal admissible extensions of the accessibility relation for belief will have to enrich the resources in terms of which the models are defined. One way to do this, a way that fits with one of the familiar strategies for responding to the Gettier counterexamples to the justified true belief analysis, is to add to the semantics for belief a theory of belief revision, and then to define knowledge as belief (or justified belief) that is stable under any potential revision by a piece of information that is in fact true. This is the defeasibility strategy followed by many of those who responded to Gettier's challenge: the idea was that the fourth condition (to be added to justified true belief) should be a requirement that there be no "defeater"—no true proposition that, if the knower learned that it was true, would lead her to give up the belief, or to be no longer justified in holding it.[18] There was much discussion in the post-Gettier literature about exactly how defeasibility should be defined, but in the context of our idealized semantic models, supplemented by a semantic version of the standard belief revision theory, a formulation of a defeasibility analysis of knowledge is straightforward. First, let me sketch the outlines of the so-called AGM theory of belief revision,[19] and then give the defeasibility analysis.

The belief revision project is to define, for each belief state (the prior belief state), a function taking a proposition (the potential new evidence) to a posterior belief state (the state that would be induced in one in the prior state by receiving that information as one's total new evidence). If belief states are represented by sets of possible worlds (the doxastically accessible worlds), and if propositions are also represented by sets of possible worlds, then the function will map one set of worlds (the prior belief set) to another (the posterior belief set), as a function of a proposition. Let **B** be the set representing the prior belief state, ϕ the potential new information, and $\mathbf{B}(\phi)$ the set representing the posterior state. Let **E** be a superset of **B** that represents the set of all possible worlds that are potential candidates to be compatible with some posterior belief state. The formal constraints on this function are then as follows: (1) $\mathbf{B}(\phi) \rightarrow \phi$ (the new information is believed in the posterior belief state induced by that information). (2) If $\phi \cap \mathbf{B}$ is nonempty, then $\mathbf{B}(\phi) = \phi \cap \mathbf{B}$ (If the new information is compatible with the prior beliefs, then nothing is given up—the new information is simply added to the prior beliefs.). (3) $\mathbf{B}(\phi)$ is nonempty if and

[17] See Schwarz & Truszczyski 1992.

[18] See Lehrer & Paxson 1969 and Swain 1974 for two examples.

[19] See Gärdenfors 1988 for a survey of the basic ideas of the AGM belief revision theory, and Grove 1988 for a semantic formulation of the theory.

only if $\phi \cap E$ is non-empty (the new information induces a consistent belief state whenever that information is compatible with the knower being in the prior belief state. and only then). (4) If $\mathbf{B}(\phi) \cap \psi$ is nonempty, then $\mathbf{B}(\phi \cap \psi) = \mathbf{B}(\phi) \cap \psi$. The fourth condition is the only one that is not straightforward. What it says is that if ψ is compatible, not with Alice's *prior* beliefs, but with the *posterior* beliefs that she would have if she learned ϕ, then what Alice should believe upon learning the *conjunction* of ϕ and ψ should be the same as what she would believe if she first learned ϕ, and then learned ψ. This condition can be seen as a generalization of condition (2), which is a modest principle of methodological conservativism (Don't give up any beliefs if your new information is compatible with everything you believe). It is also a path independence principle. The order in which Alice receives two compatible pieces of information should not matter to the ultimate belief state.[20]

To incorporate the standard belief revision theory into our models, add, for each possible world x, and for each agent i, a function that, for each proposition ϕ, takes i's belief state in x, $B_{x,i} = \{y: xD_iy\}$, to a potential posterior belief state, $B_{x,i}(\phi)$. Assume that each of these functions meets the stated conditions, where the set E, for the function $B_{x,i}$ is the set of possible worlds that are subjectively indistinguishable from x to agent i. We will also assume that if x and y are subjectively indistinguishable to i, then i's belief revision function will be the same in x as it is in y. This is to extend the positive and negative introspection assumptions to the agent's belief revision policies. Just as she knows what she believes, so she knows how she would revise her beliefs in response to unexpected information.[21]

We have added some structure to the models, but not yet used it to interpret anything in the object language that our models are interpreting. Suppose our language has just belief operators (and not knowledge operators) for our agents, and only a doxastic accessibility relation, together with the belief revision structure, in the semantics The defeasibility analysis suggests that we might add, for knower i, a knowledge operator with the following semantic rule: $K_i\phi$ is true in world x iff $B_i \phi$ is true in x, and for any proposition ψ that is true in x, $\mathbf{B}_{x,i}(\psi) \vdash \phi$. Alice knows ϕ if and only if, for any ψ that is true, she would still believe ϕ after learning ψ. Equivalently, we might define an epistemic accessibility relation in terms of the belief revision structure, and use it to interpret the knowledge operator in the standard

20 The third principle is the least secure of the principles; there are counterexamples that suggest that it should be given up. See Stalnaker 1994 for a discussion of one. The defeasibility analysis of knowledge can be given with either the full AGM belief revision theory, or with the more neutral one that gives up the fourth condition, though if we use the weaker version of the belief revision theory, the resulting logic of knowledge will be weaker: S4.2 rather than S4.3.

21 It should be noted that even with the addition of the belief revision structure to the epistemic models I have been discussing, they remain static models. A model of this kind represents only the agent's beliefs at a fixed time, together with the policies or dispositions to revise her beliefs that she has at that time. The model does not represent any actual revisions that are made when new information is actually received. The models can be enriched by adding a temporal dimension to represent the dynamics, but doing so requires that the knowledge and belief operators be time indexed, and that one is careful not to confuse belief changes that are changes of mind with belief changes that result from a change in the facts. (I may stop believing that the cat is on the mat because I learn that what I thought was the cat was the dog, or I may stop believing it because the cat gets up and leaves, and the differences between the two kinds of belief change are important.)

way. Let us say that $xR_i y$ if and only if there exists a proposition ϕ such that $\{x,y\} \subseteq \phi$ and $y \in \mathbf{B}_{x,i}(\phi)$. The constraints imposed on the function $\mathbf{B}_{x,i}$ imply that this relation will extend the doxastic accessibility relation D_i, and that it will fall between our minimal and maximal constraints on this extension. The relation will be transitive, reflexive, and strongly convergent, and so meet all the conditions of our basic theory. It will also meet an additional condition: it will be weakly connected (if xRy and xRy, then either yRz, or zRy). This defeasibility semantics will validate a logic of knowledge, S4.3, that is stronger than S4.2, but weaker than either S4F or S4.4.[22]

So, a nice, well-behaved version of our standard semantics for knowledge falls out of the defeasibility analysis, yielding a determinate account, in terms of the belief revision structure, of the way that epistemic accessibility extends doxastic accessibility. But I doubt that this is a plausible account of knowledge in general, even in our idealized setting. The analysis is not so demanding as the S4F theory, but like that theory, it threatens to let any false belief defeat too much of our knowledge, even knowledge of facts that seem unrelated. Consider the following example: Alice takes herself to know that the butler didn't do it, since she saw him in the drawing room, miles away from the scene of the crime, at the time of the murder (or so she thinks). She also takes herself to know there is zucchini planted in the garden, since the gardener always plants zucchini, and she saw the characteristic zucchini blossoms on the vines in the garden (or so she thinks). As it happens, the gardener, quite uncharacteristically, failed to plant the zucchini this year, and coincidentally, a rare weed with blossoms that resemble zucchini blossoms have sprung up in its place. But it really was the butler that Alice saw in the drawing room, just as she thought. Does the fact that her justified belief about the zucchini is false take away her knowledge about the butler? It is a fact that either it wasn't really the butler in the drawing room, or the gardener failed to plant zucchini. Were Alice to learn just this disjunctive fact, she would have no basis for deciding which of her two independent knowledge claims was the one that was mistaken. So, it seems that, on the simple defeasibility account, the disjunctive fact is a defeater. The fact that she is wrong about one of her knowledge claims seems to infect other, seemingly unrelated claims. Now it may be right that if Alice were in fact reliably informed that one of her two knowledge claims was false, without being given any information about which, she would *then* no longer know that it was the butler that she saw. But if the mere fact that the disjunction is true were enough to rob her of her knowledge about the butler, then it would seem that almost all of Alice's knowledge claims would be threatened. The defeasibility account is closer than one might have thought to the maximally demanding S4F analysis, according to which we know nothing except how things seem to us unless we are right about everything we believe.

[22] In game theoretic models, the strength of the assumption that there is common knowledge of rationality depends on what account one gives of knowledge (as well as on how one explains rationality). Some backward induction arguments, purporting to show that common knowledge of rationality suffices to determine a particular course of play (in the centipede game, or the iterated prisoners' dilemma, for example) can be shown to work with a defeasibility account of knowledge, even if they fail on a more neutral account. See Stalnaker 1996.

I think that one might plausibly defend the claim that the defeasibility analysis provides a *sufficient* condition for knowledge (in our idealized setting), and so the belief revision structure might further constrain the ways in which the doxastic accessibility relation can be extended to an epistemic accessibility relation. But it does not seem to be a plausible *necessary* and sufficient condition for knowledge. In a concluding section, I will speculate about some other features of the relation between a knower and the world that may be relevant to determining which of his true beliefs count as knowledge.

7. The Causal Dimension

What seems to be driving the kind of counterexample to the defeasibility analysis that I have considered is the fact that, on this analysis, a belief with a normal and unproblematic causal source could be defeated by the fact that some different source had delivered misinformation about some independent and irrelevant matter. Conditions were normal with respect to the explanation of Alice's beliefs about the butler's presence in the drawing room. There were no anomalous circumstances, either in her perceptual system, or in the conditions in the environment, to interfere with the normal formation of that belief. This was not the case with respect to the explanation of her belief about what was planted in the garden, but that does not seem, intuitively, to be relevant to whether her belief about the butler constituted knowledge. Perhaps the explanation of epistemic accessibility, in the case where conditions are not fully normal, and not all of the agent's beliefs are true, should focus more on the causal sources of beliefs, rather than on how agents would respond to information that they do not in fact receive. This, of course, is a strategy that played a central role in many of the responses to the Gettier challenge. I will describe a very simple model of this kind, and then mention some of the problems that arise in making the simple model even slightly more realistic.

Recall that we can formulate the basic theory of belief this way: a relation of subjective indistinguishability, for each agent, partitions the space of possibilities, and there will be a nonempty subset of each partition cell which is the set of worlds compatible with what the agent believes in the worlds in that cell. We labeled those worlds the normal one, since they are the worlds in which everything determining the agent's beliefs is functioning normally, all of the beliefs are true in those worlds, and belief and knowledge coincide. The problem was to say what the agent knows in the worlds that lie outside of the normal set. One idea is to give a more detailed account of the normal conditions in terms of the way the agent interacts with the world he knows about; we start with a crude and simple model of how this might be done. Suppose our agent receives his information from a fixed set of independent sources—different informants who send messages on which the agent's knowledge is based. The "informants" might be any kind of input channel. The agent might or might not be in a position to identify or distinguish different informants. But we assume that the informants are, in fact, independent in the sense that there may be a fault or corruption that leads one informant to send misinformation (or more generally, to be malfunctioning) while others are functioning normally. So, we might index normal conditions to the informant, as well as to the agent. For example, if there are two informants, there will be a set of worlds that is normal with respect to the input channel for informant one, and an overlapping set that is normal for informant two. Possible worlds in which conditions are fully normal will be those in which all the input channels are functioning normally—the worlds in the intersection of all of the sets.[23] This intersection will be the set compatible with the agent's beliefs, the set where belief and knowledge coincide. If conditions are abnormal with respect to informant one (if that information channel is corrupted) then while that informant may influence the agent's beliefs, it won't provide any knowledge. But if the other channel is uncorrupted, the beliefs that have it as their sole source will be knowledge. The formal model suggested by this picture is a simple and straightforward generalization of the S4F model, the maximal admissible extension of the doxastic accessibility relation. Here is a definition of the epistemic accessibility relation for the S4F semantics, where $\mathbf{E}(x)$ is the set of worlds subjectively indistinguishable from x (to the agent in question) and $\mathbf{N}(x)$ is the subset of that set where conditions are normal (the worlds compatible with what the agent believes in world x): $x\mathrm{R}y$ if and only if $x \in \mathbf{N}(x)$ and $y \in \mathbf{N}(x)$, or $x \notin \mathbf{N}(x)$ and $y \in \mathbf{E}(x)$. In the generalization, there is a finite set of normal-conditions properties, $\mathbf{N}^j$, one for each informant j, that each determines a subset of $\mathbf{E}(x)$, $\mathbf{N}^j(x)$, where conditions are functioning normally in the relation between that informant and the agent. The definition of R will say that the analogue of the S4F condition holds for each $\mathbf{N}^j$. The resulting logic (assuming that the number of independent information channels or informants is unspecified) will be the same as the basic theory: S4.2.

Everything goes smoothly if we assume that information comes from discrete sources, even if the agent does not identify or distinguish the sources. Even when the agent makes inferences from beliefs derived from multiple sources, some of which may be corrupt and others not, the model will determine which of his true beliefs count as knowledge, and which do not. But in even a slightly more realistic model, the causal explanations for our beliefs will be more complex, with different sources not wholly independent, and deviations from normal conditions hard to isolate. Beliefs may have multiple interacting sources—there will be cases of overdetermination and preemption. There will be problems about how to treat cases where a defect in the system results, not in the reception of misinformation, but from the failure to receive a message. It might be that had the system been functioning normally, I would have received information that would have led me to give up a true belief. And along with complicating the causal story, one might combine this kind of model with a belief revision structure, allowing one to explore the relation between beliefs about causal structure and policies for belief revision, and to clarify the relation between the defeasibility analysis and an account based on the causal strategy. The abstract problems that arise when one tries to capture a more complex structure will reflect, and perhaps help to clarify, some of the patterns in the counterexamples that arose in the post-Gettier literature. Our simple model avoids most of these problems, but it is a start that may help to provide a context for addressing them.

[23] It will be required that the intersection of all the normal-conditions sets be nonempty.

Appendix

To give a concise summary of all the logics of knowledge I have discussed, and their corresponding semantics, I will list, first the alternative constraints on the accessibility relation, and then the alternative axioms. Then I will distinguish the different logics, and the semantic conditions that are appropriate to them in terms of the items on the lists.

Conditions on the accessibility relation

(Ref)	$(x)xRx$
(Tr)	$(x)(y)(z)((xRy \,\&\, yRz) \rightarrow xRz)$
(Cv)	$(x)(y)(z)((xRy \,\&\, xRz) \rightarrow (\exists w)(yRw \,\&\, zRw))$
(SCv)	$(x)(\exists z)(y)(xRy \rightarrow yRz)$
(WCt)	$(x)(y)(z)((xRy \,\&\, xRz) \rightarrow (yRz \vee zRy))$
(F)	$(x)(y)(xRy \rightarrow ((z)(xRz \rightarrow yRz) \vee (z)(xRz \rightarrow zRy))$
(TB)	$(x)(y)((xRy \,\&\, x \neq y) \rightarrow (z)(xRz \rightarrow zRy))$
(E)	$(x)(y)(z)((xRy \,\&\, xRz) \rightarrow yRz)$

Axioms of the different systems

(T)	$K\phi \rightarrow \phi$
(4)	$K\phi \rightarrow KK\phi$
(4.2)	$MK\phi \rightarrow KM\phi$
(4.3)	$(K(\phi \rightarrow M\psi) \vee K(\psi \rightarrow M\phi))$
(f)	$((M\phi \,\&\, MK\psi) \rightarrow K(M\phi \vee \psi))$
(4.4)	$((\phi \,\&\, MK\psi) \rightarrow K(\phi \vee \psi))$
(5)	$M\phi \rightarrow KM\phi$

The logics for knowledge we have considered, and the corresponding semantic constraints on R relative to which they are sound and complete, are as follows: The logics are of increasing order of strength, the theorems of each including those of the previous logics on the list.

S4	K + T + 4	Ref + Tr		
S4.2	S4 + 4.2	Ref + Tr + SCv	OR	Ref + Tr + Cv
S4.3	S4 + 4.3	Ref + Tr + SCv + WCt	OR	Ref + Tr + WCt
S4F	S4 + f	Ref + Tr + F		
S4.4	S4 + 4.4	Ref + Tr + TB		
S5	S4 + 5	Ref + Tr + E		

In each of the logics of knowledge we have considered, from S4.2 to S4.4, the derived logic of belief, with belief defined by the complex operator MK, will be KD45. (In S4, belief is not definable, since in that logic, the complex operator MK does not satisfy the K axiom, and so is not a normal modal operator. In S5, belief and knowledge coincide, so the logic of belief is S5.) KD45 is K + D + 4 + 5, where D is $(K\phi \rightarrow M\phi)$. The semantic constraints are Tr + E + the requirement that the accessibility relation be *serial*: $(x)(\exists y)xRy$.

In a semantic model with multiple knowers, we can add a common knowledge operator, with an accessibility relation defined as the transitive closure of the epistemic accessibility relations for the different knowers. For any of the logics, from S4 to S4.4, with the corresponding semantic conditions, the logic of *common* knowledge will be S4, and the accessibility relation will be transitive and reflexive, but will not necessarily have any of the stronger properties. If the logic of knowledge is S5, then the logic of common knowledge will also be S5, and the accessibility relation will be an equivalence relation.

2

Luminosity and the KK Thesis

1. Introduction

"The mind," according to a Cartesian picture, "is transparent to itself. It is of the essence of mental entities, of whatever kind, to be conscious, where a mental entity's being conscious involves its revealing its existence and nature to its possessor in an immediate way. This conception involves a strong form of the doctrine that mental entities are 'self-intimating,' and usually goes with a strong form of the view that judgments about our own mental states are incorrigible or infallible, expressing a super-certain kind of knowledge which is suited for being an epistemological foundation for the rest of what we know."[1] This is Sydney Shoemaker's characterization of a picture that he, along with most epistemologists and philosophers of mind of the twentieth century, rejected. Timothy Williamson, writing at the end of the twentieth century, summed up the general idea of the Cartesian picture this way: "There is a constant temptation in philosophy to postulate a realm of phenomena in which nothing is hidden from us, . . . a cognitive home in which everything lies open to our view."[2] These impressionistic descriptions obviously need to be pinned down to a specific thesis if one is to give an argument against the Cartesian picture of the mind and of our knowledge of its contents, and different anti-Cartesian philosophers have done this in different ways. At least some critics of the picture, including Shoemaker, want to allow for distinctive epistemic relations of some kind that a subject bears to his or her mind—to one's own experiences and thoughts—even while rejecting the kind of transparency that Shoemaker's and Williamson's Cartesian assumes. The challenge is to spell out, explain, and defend the distinctive kind of epistemic relations.

Shoemaker, in the lectures from which this characterization of the Cartesian picture is taken, focuses on the notion of introspection, and his main target was not the Cartesian picture, but what he called "the perceptual model" of introspection, a view that holds that "the existence of mental entities and mental facts is, logically speaking, as independent of our knowing about them introspectively as the existence of physical entities and physical facts is of our knowing about them perceptually." His aim was to find a middle ground between the Cartesian view and this perceptual model, a view according to which there are constitutive conceptual connections between our experience and thought and our introspective knowledge of it. He

[1] Shoemaker 1994, 271. [2] Williamson 2000, 93.

argues, for example, that a certain kind of "self-blindness," a condition in which a rational, reflective, and conceptually competent agent is ignorant of his or her beliefs, is incoherent.

While there is a sense in which Shoemaker's own view falls between the Cartesian picture and the perceptual model, there is also a way to understand the Cartesian view so that it is a version of the perceptual model. One can interpret the Cartesian as holding that it is right to say that we perceive the contents of our minds, which are conceptually independent of our perceiving them, but we do so with a special perceptual capacity that is direct and infallible. Whether or not this is the right way to think of the Cartesian view, it does seem that some versions of this picture at least exploit the metaphors of perception to describe the way we know what we feel and think. For example, in saying that we are directly *acquainted* with the contents of our mind, the Cartesian is co-opting what is, in its more ordinary use, a causal relation between distinct things. Williamson's choice of the term "luminous" for characterizing the kind of access to the mind that the Cartesian tempts us to postulate also suggests this way of thinking of the Cartesian picture: a luminous state (according to the metaphor) is one that emits a special light that renders it essentially perceptible. But Williamson does not rely on the metaphor: he defines luminosity in more sober terms: a state or condition is *luminous* if and only if a person who is in that state or condition is thereby in a position to know that he or she is in it. Williamson's rejection of the Cartesian picture is more uncompromising that Shoemaker's, whose rejection of the perceptual model allows for mental states that are luminous in this sense. But Williamson argues that there are no nontrivial states or conditions that are luminous.

Williamson's general picture of knowledge and the mind is thoroughly externalist. "Externalism" (like the contrasting "Cartesian picture") can be pinned down in various ways, but the general idea is that we should understand subjects—those who are able to experience, think, and know about the world—from the outside.[3] We should formulate the philosophical questions about knowledge and intentionality as questions about the relations that hold between one kind of object in the world (those capable of experience and thought) and the environments they find themselves in. More specific externalist theses, such as reliabilist accounts of knowledge, and anti-individualist accounts of the intentional content of thought, are developed in the context of such a general picture. There are tensions between the externalist theses and the possibility of mental states that are luminous in Williamson's sense, but there are many attempts to resolve the tensions. Tyler Burge, for example, argues that his anti-individualism about mental content is compatible with the thesis (to put it roughly) that a thinker knows the content of his or her thought in virtue of thinking it.[4] The general strategy of those who aim to reconcile an externalist, anti-Cartesian conception

[3] See chapter 1 of Stalnaker 2008 for my attempt to characterize the general externalist perspective. In chapter 6 of that book, I address the tension between an externalist account of mental content and the thesis that the content of a thinker's thoughts is, in a sense transparent to the thinker. Though I don't put it this way, one can see the arguments of that chapter as an attempt to reconcile a certain kind of luminosity with an externalist and anti-Cartesian picture.

[4] Burge 1988.

of knowledge and the mind with a limited kind of luminosity is to argue, as Shoemaker does, for constitutive conceptual connections between some mental states and our knowledge of them. Selim Berker follows this strategy in his critique of Williamson's anti-luminosity argument. His claim is that Williamson "presupposes that there does not exist a constitutive connection between the obtaining of a given fact and our *beliefs* about the obtaining of those facts," and that his anti-luminosity argument succeeds only with this presupposition.[5] I agree with this general strategy, but one has to look at the details.

My plan in this chapter is first to sketch Williamson's master argument against the possibility of luminous states or conditions, looking first at a version of the argument aimed at the thesis that phenomenal states are luminous, and then at a version applied to knowledge itself: an argument against the thesis that *x knows that P* implies that *x knows that x knows that P* (or at least that knowing that P puts one in a position to know that one knows that P). While I have some sympathy with the argument as applied to phenomenal states, I will try to show that the refutation of the KK thesis does not work. I will argue this by giving a simplified and idealized model of knowledge—one that is thoroughly externalist—in which the KK thesis holds, and then considering where the anti-luminosity argument goes wrong for that idealized concept of knowledge. While the model I will define is artificial, I think it reflects some essential features of the notion of knowledge that we apply to ourselves, and that is of concern in epistemology and the philosophy of mind.

2. Safety and Margins of Error

Williamson's anti-luminosity arguments are given for particular examples of concepts that are the most favorable cases for luminosity, but the arguments all take the same generalizable form, and they are all based on a "margin-of-error" principle, which is in turn motivated by a safety principle for the concept of knowledge. The safety principle is that knowledge that ϕ implies, not only that ϕ is true, but that it is *safely* true. Safety, as Williamson explains it, is one of a family of concepts that include reliability, robustness, and stability. He illustrates the general notion with an example of a contrast between a ball balanced on the tip of a cone in a state of *unstable* equilibrium, and a ball sitting at the bottom of a hole in a state of *stable* equilibrium. In the former case, the ball falls out of equilibrium in "nearby" possible situations where, for example, the surrounding air currents are very slightly different, while in the latter case minor variations in the environment would have no such effect.[6] Safety for knowledge implies that the truth of the proposition known must be stable in this sense: true not only in the actual situation, but also in all "nearby" possible situations. But the unstable equilibrium example brings out the fact that the kind of "nearness" of possible situations that is relevant to a safety condition for knowledge is distinctive, and not exactly the same as the kind of "nearness" that is relevant to the equilibrium example. One observing the ball in unstable equilibrium can know that it is not falling since even if the ball is not safe from falling, the

[5] Berker 2008. [6] Williamson 2000, 123.

observer is safe from falsely believing that the ball is falling. This is because if it were to fall, the observer would see that it did, and so no longer believe that it is not falling. Williamson puts the safety condition this way: "In case α one is safe from error in believing that C obtains if and only if there is no case close to α in which one falsely believes that C obtains."[7]

The anti-luminosity arguments do not give any general characterization of the notion of nearness of possible situations that is relevant to safety for knowledge. Instead, they describe for the particular examples a sequence of possible situations, each of which is assumed to be similar in the relevant respects to its neighbors. The specific "margin of error" premises of the anti-luminosity arguments are motivated by the idea of safety, but they are stated for the particular example for which the argument is given.

3. Phenomenal States

Williamson begins with the example of the condition that one *feels cold*. He imagines a person who feels freezing cold at dawn, and then slowly warms up until, at noon, he feels hot. It is supposed, plausibly, "that one's feelings of heat and cold change so slowly during this process that one is not aware of any change in them over one millisecond." He then considers a sequence of times, one millisecond apart, from dawn until noon. If t_i is one of the times, then α_i is the situation of the subject at t_i. The margin of error principle, for this example, is that for each time t_i,

(I_i) If in α_i one knows that one feels cold, then in α_{i+1} one feels cold.[8]

If we assume, for reductio, that "feeling cold" is luminous—that one who feels cold is thereby in a position to know that he feels cold—and also assume that the person is throughout the time period actively considering whether he feels cold, we will have for each time t_i,

(II_i) If in α_i one feels cold, then in α_i one knows that one feels cold.

From all of the premises of these forms, one can derive that the subject still feels cold at noon, contrary to the stipulation that he then feels hot. To break the chain of inferences to this conclusion, we must assume that at least one of the premises (II_i), is false, which implies that "feeling cold" is not a luminous condition.

More cautiously, we might say that one must reject either one of the premises (II_i) or one of the premises (I_i). If there is a tight enough conceptual connection between feeling cold and believing that one feels cold, then one might assume that the point at which one stops feeling cold is, more or less by definition, the same as the point at which one stops believing it. If this is true, then Williamson's safety condition might be satisfied, even though one of the margin-of-error premises of the argument is false. Let α_i be the last point in the sequence at which the subject feels cold, and also the last point in the sequence at which the subject believes that he feels cold. So, at α_{i+1}, the subject neither feels cold nor believes that he does. If we assume that at α_i,

[7] Ibid., 126–7. [8] Ibid., 97.

he knows that he feels cold, then one of the margin-of-error premises is false, but there is no violation of the safety condition, since in the very similar situations in which it is false that the subject feels cold, he no longer believes it. So, the case is like the case where the ball in unstable equilibrium falls out of equilibrium in very similar situations, even though this does not prevent the observer from knowing that it is in equilibrium, since she would no longer believe it was in equilibrium in those nearby situations. (This is Berker's diagnosis.)

One might try to avoid this problem by reformulating the safety condition for knowledge in a way that avoids mentioning belief. Let us say that two types of states or conditions that a person could be in are *robustly indistinguishable* to the person if the following is true: Were an experimenter to put the subject into one of the states, and then shift it to the other, and back, the subject, even when attending carefully, would be completely unable to tell when the shifts took place. The subject might be told, when in one of the states that it is state A, and when in the other that it is state B, and then do no better than chance at identifying one of the two states as A, or as B. Using a notion like this, one might characterize a safety condition as follows:

> In case α one is safe from error in believing that C obtains if and only if there is no case β close to α, and robustly indistinguishable from it, such that C does not obtain in β.

Since, in defining the series of times in the anti-luminosity argument for feeling cold, Williamson specifies that the times, one millisecond apart, are indistinguishable in something like this sense, it seems that this version of a safety requirement would suffice to imply the margin of error premises.

We have bypassed the notion of belief in stating our safety condition, but it seems that we might also bypass belief in stating a thesis that is motivated by the plausible assumption that there is some kind of constitutive connection between being in a phenomenal state like feeling cold and being aware that one is. It seems reasonable to hold that if states A and B are robustly indistinguishable then, if the subject feels cold in state A, then he or she also feels cold in state B. But if one accepts this assumption, then the kind of sorites sequence that Williamson uses to argue against the luminosity of the state of feeling cold will pose a problem that is independent of any assumptions about knowledge, safety, or margins of error. Our assumption will license the following variations on the luminosity premises:

> (I^*_i) If in α_i one feels cold, then in α_{i+1} one feels cold.

Given that the subject feels cold at dawn, and not at noon, this is enough for a contradiction.

This argument is a version of what Michael Dummett called "Wang's paradox."[9] In his discussion of this paradox, Dummett considered both phenomenal concepts and what he called "observational predicates" such as being red or green. If the relevant predicates apply to things in the world, as "red" and "green" do, then one may respond to the paradox by denying the indistinguishability assumption for such

[9] Dummett 1975.

predicates. I think it is reasonable to hold that even if observation plays a role in fixing the reference of color terms, it is still possible for things to differ in color in cases where observers cannot tell them apart. But it is harder to make this move in cases where the predicate characterizes a phenomenal state, since it is at least arguable that distinct but subjectively indistinguishable states cannot be phenomenal states. Dummett took one upshot of the argument to be that "there are no phenomenal qualities, as these have been traditionally understood."[10]

Wang's paradox is of course a version of the sorites paradox, which is a problem about vagueness. While Williamson agrees that states such as *feeling cold* are vague, he argues that his anti-luminosity arguments are not dependent on the vagueness of the relevant predicates, and not dependent on his own controversial epistemic account of vagueness. Specifically, he argues that the anti-luminosity argument will still work even for artificially sharpened versions of the predicates for such states. But it matters how one sharpens the predicates. Berker's argument is that there may be a constitutive conceptual connection between *feeling cold* and *believing that one feels cold*, and a connection of this kind seems to motivate the thesis that situations indistinguishable to the subject cannot differ with respect to whether he feels cold. If there is a connection of this kind between these two vague notions, than one must be sure to sharpen them together in a way that preserves the connection. Williamson suggests that we might "sharpen 'feels cold' by using a physiological condition to resolve borderline cases."[11] But this would be to leave out a constitutive connection between *feeling cold* and any doxastic state, and I think one should agree, independently of the sorites-style anti-luminosity argument, that states explained this way are not luminous.

Here is one kind of story one might tell about the acquisition and character of phenomenal concepts for states such as *feeling cold*,[12] a story on which Williamson's way of sharpening the vague concept would be appropriate: First, one learns that under certain favorable conditions one is able to tell, by the way one feels, that one is in a cold environment. ("Cold" here is both vague and context-dependent, but it refers to an objective state of the environment.) One then hypothesizes that there is an internal state-type that one is in when one is inclined to report, on the basis of how one feels, that it is cold in one's environment—a state that one calls "*feeling cold*." This hypothesized state is what explains one's ability to tell, by the way one feels, something about the temperature of the environment. Sometimes one judges that one is in the hypothesized internal state even when one knows that it is not actually cold in the environment, and one also recognizes that one may fail to feel cold, even when it is in fact cold. Scientists might confirm the hypothesis by identifying a state of the nervous system that explains one's capacity to know, in the right conditions, about the temperature in the environment. It is this physiological state (according to this way of thinking about phenomenal states) that is the realization of the functional property of feeling cold.

On this kind of story, the inclination to judge that one's environment is in a certain state plays a reference-fixing role in determining the relevant internal state,

[10] Ibid., 232. [11] Williamson 2000, 103.
[12] Wilfred Sellars told a story like this in Sellars 1956/1997.

but the state itself has no necessary connection, either with the actual state of the environment, or with the inclination to form a belief about it. The story suggests that the following might happen: one goes to the doctor with the complaint that one is always feeling cold, even when the weather is unusually warm. After a thorough neurological examination, the doctor reports that you do not in fact feel cold in these situations: you are suffering from a persistent illusion that you do. Or alternatively, you go to the doctor with a more mystifying complaint: you find that you can no longer tell how cold it is. You report that if you go out in your shirtsleeves on a frigid winter day, it feels just as warm as when you were inside before the fire. After a thorough neurological examination, the doctor reports that the good news is that you *do* in fact feel cold when you go out. The bad news is that you suffer from a condition that blocks your cognitive access to this feeling so that you can't tell that you have it.

Even granting that the doctor has got the physiology right, you might be inclined to resist her description of your situation, in either of these scenarios. You might insist (in the first case) that even if the explanation for your feeling is an abnormal one, if it seems to you that you feel a certain way, then you do. In the second case, you might insist that you don't have the feeling if you don't feel it, and you don't feel it if you are not at least in a position to recognize that you are feeling it. But the phenomenology may be complicated in abnormal cases, and the subject may be unsure or ambivalent about how such situations should be described.

Suppose that instead of following Williamson's suggestion to sharpen artificially the notion of *feeling cold* by identifying it with a physiological condition, we sharpen it in a way that preserves a constitutive doxastic connection. Suppose we take *feeling cold* to be something like the state of being inclined to believe, based on how one feels, that one is in a cold environment. Now what happens when one goes through the sorites sequence? It is reasonable to suppose, as Williamson does, that the subject's answers to the question, "do you feel cold?" will be firmly positive at the start, but at a certain point will begin to be hesitant and qualified, perhaps with some backtracking ("I guess I didn't really feel cold a moment ago either"). Suppose one sharpens the concept by saying that at the first sign of hesitation, the subject no longer qualifies as feeling cold, since he is no longer inclined to report, unequivocally, that he feels cold. At that same point, he no longer counts as believing that he feels cold. There is some distortion in this sharpening, and in the assumption of a tight link between the phenomenal state and the disposition to report, since the hesitation and qualification might take an epistemic form ("I'm not quite sure whether I still feel cold"), and the backtracking suggests that the subject is allowing for the possibility that his judgments about whether he feels cold might be mistaken. Furthermore, if we sharpen our revisionary stipulation about what it means to feel cold, and to believe it, in this way, we can expect that there will be some randomness in the determination of the exact point where hesitation begins. There will be points at which the subject does not have a stable *disposition* to report that he feels cold (as contrasted with a more hesitant answer). Sometimes, when in an otherwise indistinguishable state, he will express hesitation, and sometimes not.

It seems that either way of sharpening the vague phenomenal concept will yield an artificial concept that lacks something that seems essential to phenomenal concepts. The explanation may be that there is some incoherence in a notion of a phenomenal state that combined an essentially epistemic or doxastic dimension with the assumption that the state is an intrinsic, nonintentional internal state that one is judging oneself to be in when one reports that one feels cold, or is in pain, or is having a sensation of red. The vagueness of such phenomenal concepts helps to obscure the incoherence, and the attempt to sharpen them artificially brings the incoherence to the surface. If something like this is right, then phenomenal concepts do not offer the best place to challenge Williamson's anti-luminosity thesis and argument. But purely epistemic and doxastic concepts may be better cases.

4. In Defense of the KK Principle

The KK principle, in its unvarnished form, says that the proposition *that x knows that P* entails the proposition *that x knows that x knows that P*. Jaacko Hintikka endorsed this principle in his early formulation of epistemic logic,[13] and it is accepted in many developments and applications of this kind of formal semantic theory. As acknowledged from the start, by Hintikka and others, epistemic logic is an idealized and schematic model of knowledge, and not a realistic description. The most that could plausibly be said about a notion of knowledge that applies to real human agents is that knowledge that P puts one in a position to know that one knows that P. But this is enough to imply that knowledge is a luminous concept, in Williamson's sense, and Williamson applies his signature anti-luminosity argument to rebut the thesis that the principle holds. I will question the soundness of the argument by questioning the relevant margin of error principles, and also the more general safety condition on knowledge. I will begin with a simple model of knowledge that validates the KK principle, and a simple example to illustrate it. After seeing where the anti-luminosity argument breaks down for this simple case, I will consider some of the similarities and differences between the simple model and a notion of knowledge that might apply to ourselves.

Knowledge, whatever else it may be, is a state of registering or carrying information, where information is understood in terms that Fred Dretske, among others, has developed.[14] According to the general picture, an object or system registers information about some aspect of its environment if it is causally sensitive in a systematic way to a range of facts. Specifically, a system functions to carry information if it is capable of being in a range of internal states, $S_1, \ldots S_n$ which correspond to a range of states of the environment, $E_1, \ldots E_n$ in the following way: under certain favorable conditions, the object is in states S_i, for each i, if and only if the environment is in the corresponding state E_i. The object or system then carries the information that ϕ if and only if the conditions are favorable in the relevant sense, it is in state S_i, and the environment being in states E_i entails that ϕ. Strictly, the object or system "knows" or carries the information that ϕ only if ϕ is a proposition that distinguishes between the possible situations that satisfy the favorable conditions. Trivially, propositions that

[13] Hintikka 1962.

[14] The classic development of this way of understanding knowledge is Dretske 1981.

state those conditions will be propositions that are entailed by all of the states E_i, but these propositions are presupposed by the system, or by the interpreter of the system.

This abstract framework can allow for misinformation by removing the qualification that the conditions are favorable in the relevant sense. That is, a system that functions to carry information might be said to "believe" that ϕ whether or not conditions are favorable, provided it is in a state S_i that is such that, if conditions were favorable, it would carry the information that ϕ.

This simple story will apply to a wide range of objects and systems that are not rational agents—to artifacts such as thermostats and fuel gauges, to impersonal and subpersonal biological systems such as immune systems, and to suprapersonal entities such as economic systems. And there are even simpler cases of information registration: To repeat some often cited examples, the reflection on the lake carries information about the sky above, and the rings on a tree trunk carry information about the age of the tree. But even if information registration is ubiquitous, manifested in things very different from the beings that are of interest to epistemologists, we do naturally use epistemological language metaphorically to describe simpler cases of information carrying. A coil that returns to a certain shape after being deformed has a "memory," since it retains the information about its default shape. The immune system can "recognize" foreign invaders, though it may also "mistake" the body's own tissue for a foreign invader (in cases of auto-immune disease). These metaphors reflect the fact that the simple cases share an abstract structure with the kind of complex and sophisticated information-carrying devices that rational agents are, and I think it helps to understand our own knowledge and belief to start with simpler cases that share the structure, and that may face simpler versions of the epistemological problems that human knowers face. The appeal to this abstract structure to throw some light on philosophical problems about knowledge and intentionality need not be part of any reductive project, either the search for an analytic definition of knowledge, or an attempt to give a materialist reduction of intentionality.

So, my examination of the anti-luminosity argument against the KK principle will begin with a simple artifact whose function is to carry information—a measuring device.[15] I want my device to be understood, not just as something that provides information to its users, but as itself a knower, or a simple model of a knower. My device is a digital kitchen scale that I use to weigh the coffee beans before I grind them for our morning coffee. It weighs them in grams, and my target weight is 42 grams. Sometimes, when I am thinking about Williamson's anti-luminosity argument, I add the beans slowly, one at a time, after the scale registers 41. At a certain point sometimes it will flicker a bit between 41 and 42, and a bean or two later will settle in at 42. I don't then take my scale to "know" that the bowl of coffee beans weighs *exactly* 42 grams—it is not that accurate. But it is pretty accurate, so I think I can reasonably assume that it "knows" that the beans weigh 42 ± 1 grams. What this "margin of error" assumption shows, in terms of the abstract structure described, is

[15] My discussion is influenced by unpublished work by Damien Rochford, who develops the analogy between rational agents and systems of measuring instruments in illuminating detail.

that while the alternative states of the information-carrying device, $S_1, \ldots S_n$ (the numbers registered on the scale's screen), are pairwise disjoint states, the corresponding states of the external environment (the ranges of weights) need not be disjoint. So, for example, S_{42} and S_{43} (42 and 43 on the screen) will correspond to the ranges 41–43 and 42–44 respectively for the weight in grams of the beans. But taking account of *all* the information carried by a state of the system, the alternative states of the world corresponding to the states *will* be disjoint. This is because the state S_i trivially carries information about itself—specifically, that it is in state S_i. It will always be true that the scale reads 42 only if it reads 42, under favorable (or unfavorable) conditions. So, the state of the world corresponding to S_{42} (under favorable conditions) will the state in which the scale is in state S_{42}, and the actual weight is in the range 41–43 grams.

The description of the abstract structure includes the crucial qualification "under certain favorable conditions," and we pin down a particular application of the information-carrying story only when we specify (explicitly or implicitly) those conditions (*channel conditions*, to use an information-theoretic term, or what Dennis Stampe called "fidelity conditions" in an early paper on this kind of approach to intentional content[16]). In general, the channel conditions will include two different kinds of restriction: first, the correlation will be required to hold only conditional on the proper functioning of the internal workings of the information-carrying device; second, the correlation is conditional on certain features of the external environment. My scale, if it malfunctions, may register a number that fails to correspond, even within the margin of error, to the actual weight. Perhaps because of a low battery it sticks on a certain number, or perhaps the zero point is off so that weights are systematically too high by two or three grams. But the registration might also diverge from the actual value because some anomalous atmospheric or gravitational phenomenon creates an environment in which even a well-functioning scale will fail to give an accurate measurement. Maybe it is steel balls rather than coffee beans we are weighing, and there is a magnet below the scale that distorts the measurement. The scale "knows" the weight of what is on the scale only when the channel conditions of both kinds are satisfied. These conditions will be constitutively connected with the margin of error that fixes the specific content of the information that is registered. If the channel conditions are such that they may be satisfied when the actual weight is 41 and the scale registers 42, then this must be reflected in the specification of the content of the information that the number 42 registers, where this specification is given by the margin of error. And if under the specified channel conditions, the scale measures 42 *only* when the actual weight is between 41 and 43, then the measurement carries the information that the actual weight is within that range.

While the channel conditions and the margin of error are tightly connected, there is some flexibility, and perhaps some arbitrariness, in exactly what the conditions and the corresponding margin of error each are. The line between normal fluctuation and malfunction might be drawn, by an interpreter of a system that functions to carry information, in alternative ways. Some differences between an internal malfunction

[16] Stampe 1977.

and the proper functioning of a device may be relatively sharp and inflexible, but at least some of the channel conditions may involve continuities that will require arbitrary cutoffs, at least if we are to avoid vagueness. Consider a slightly different measuring device, an oven thermometer that is designed to measure the average temperature inside the oven. We can be reasonably sure that the molecular energy within the oven is approximately uniform: the temperature of the air in the small part of the space near the sensor will normally be close to the average temperature within the whole oven. But there is a small probability that it will be very different, and when it is, the thermometer will sometimes register a temperature that diverges significantly from the average temperature that it is ostensibly measuring. Thermodynamics will tell us, for each margin of error, how probable it will be that the difference between the average temperature and the temperature in the space close to the sensor will remain within that margin. But how improbable must a distribution be to count as abnormal? The answer to that question will contribute to determining the margin of error, but we should expect the right answer to be both vague and context-dependent. As with the infamous bank cases, used to motivate contextualist and subject-sensitive invariantist accounts of knowledge, this kind of choice may be constrained by the decision-making role that the information-registering device is playing, and by the importance of accuracy for making the right decision.

But who sets these numbers—who specifies the channel conditions, and the resulting margins of error? Remember, we are thinking of our devices, not as things we use to acquire knowledge, but as themselves models of knowers. But knowers (real ones, as well as such artificial models) are also devices that other knowers use to acquire knowledge, and it is the attributors of knowledge (including epistemological theorists) who specify, or presuppose, the conditions that determine the content of the knowledge attributed. The information-theoretic story about knowledge is a thoroughly external one that characterizes the state from the outside, as a feature of things in the world that are sensitive to facts about their environment. Sensitivity to facts about the environment is essentially contrastive: knowers represent things as being *this* way rather than *that* way, and what the contrast is will depend on the background facts that are presupposed by the attributor. There is no escaping the context-dependence of such attributions (on this way of understanding knowledge), since things carry information only relative to facts about the causal relations between those things and other things in their world. They don't carry information *about* those facts, but carry information in virtue of the facts being the way they are. This dependence of information registration on a factual background will be uncontroversial for our simple measuring devices, but I think it is also an unavoidable feature of the more sophisticated measuring devices that we ourselves are.[17]

I want now to look at a version of Williamson's argument against the KK principle, applied to the schematic account I have given of a system that functions to carry information. His argument is a reductio of the KK principle, and we know that it

[17] I used the information story about knowledge in Stalnaker 1993 to give a theoretical rationale both for an anti-individualist account of the content of knowledge and belief, and for an essentially contextualist account of knowledge.

cannot be sound when applied to this schematic account, since the KK principle holds on this account, and it will be clear exactly where the argument goes wrong. Just to be clear on the dialectic of my argument: The critique of this application of the anti-luminosity argument should not be controversial, and does not, by itself, have force against Williamson's anti-KK argument as applied to the ordinary notion of knowledge that is his target. The controversy will be about the extent to which the structure of the schematic account of systems like measuring devices that function to carry information is reflected in an adequate account of real knowledge. But the critique of this variant of the argument does accomplish two things: First, it illustrates the fact that there is a thoroughly externalist, anti-Cartesian conception of something like knowledge that allows for luminosity, in Williamson's sense. Second, by pinpointing the place at which this version of the argument fails, it focuses attention on the corresponding point in Williamson's actual argument, so that we can consider whether the differences between simple systems that function to carry information and real knowers makes a difference for the soundness of the argument.

As with the anti-luminosity argument applied to phenomenal states like *feeling cold*, the central premises of Williamson's argument against the KK thesis are margin-of-error (ME) principles that say that one can know that ϕ in situation α only if ϕ is true, not only in α, but also in situations that are very similar to α. Our kitchen scale weighs things with a margin of error $\pm$ 1 gram, so we can assume that two situations where the item being weighed differs in weight by less than this margin of error are sufficiently similar for the purposes of Williamson's ME premises. So, consider a sequence of propositions, ϕ_0, ϕ_1, $\phi_2, \ldots$ defined as follows: ϕ_i is the proposition that the beans in the bowl weigh no less than $40 + i \times (.1)$. For example, ϕ_0 is the proposition that the weight of the beans $\geq$ 40 grams, and ϕ_{23} is the proposition that the weight of the beans in the bowl $\geq$ 42.3 grams. The number .1 is well within the margin of error, so the ME principles for this example are the sentences of the following form:

$$(K\phi_i \rightarrow \phi_{i+1}).$$

(If it is *known* that the beans weigh no less than x, then it is *true* that they weigh no less than x+.1.)

Williamson's argument assumes that the margin-of-error principles are not only true, but also known by the knower. Unlike Mr. Magoo (the estimator of the height of a tree in the example used in Williamson's actual argument), our humble scale is not a reflective agent, but the knowledge operator, as interpreted in the semantics for our schematic model, represents what is true in all possible situations compatible with the information carried by the scale. So, it includes the channel conditions that are presupposed, as well as other features of the system that are presupposed to hold. So, let us assume that if the ME principles are true, then they are "known" in this sense. So, we have

$$(\text{ME})K(K\phi_i \rightarrow \phi_{i+1}).$$

Along with these premises, we have the KK principle

$$(\text{KK})(K\phi \rightarrow KK\phi)$$

For Williamson, the KK principle is a hypothesis assumed for a reductio, but in our model the KK principle is true. Here is an argument for this claim: Suppose the system is in a state S_i such that the corresponding state is E_i, and that E_i is a state that obtains (assuming that the channel conditions are satisfied) only if ϕ. Suppose further that $K\phi$ is true, which presupposes that the channel conditions obtain. Then since E_i is a state in which S_i obtains, it follows that E_i obtains (assuming that the channel conditions are satisfied) only if $K\phi$. So, since the channel conditions do obtain, it follows that $KK\phi$ is true.

Now here is the reductio argument, modeled on Williamson's argument. Since the KK principle holds, this will be a reductio of one of the ME premises.

Assume that the scale registers 42, and that channel conditions are satisfied, so that the scale "knows" that the weight is between 41 and 43, inclusive. It follows that it "knows" that the weight is no less than 40, so we have

(1) $K\phi_0$.

It also follows that it "knows" that the actual weight is *not* greater than 43, from which it follows that it is false that the weight is no less than 43.1, and so false that the scale "knows" this. So, we have

(2) $\sim K\phi_{31}$.

Now from each of the (ME) premises, $K(K\phi_i \rightarrow \phi_{i+1})$, we get, by distributing the K,

(3) $KK\phi_i \rightarrow K\phi_{i+1}$

(Following Williamson, and standard epistemic logic, we are assuming deductive closure for knowledge.)

From (3) and (KK), we deduce, by the transitivity of the material conditional arrow,

(4) $K\phi_i \rightarrow K\phi_{i+1}$

for each *i*. Then by a sequence of 31 modus ponens steps, using 31 instances of (4), and starting with (1), we get

(5) $K\phi_{31}$

which contradicts (2).

The argument is obviously valid, so which of the premises is false? Since (KK) and deductive closure for K are true in this simple idealized model, and since the stipulations of the case ensure that premises (1) and (2) are true, we know that at least one instance of (ME) is false. It is easy to see that the culprit is $K(K\phi_{10} \rightarrow \phi_{11})$. Our assumptions about the case imply that the scale knows that the weight is no less than 41, but it might be, for all it knows, that the weight is *exactly* 41, in which case ϕ_{11} would be false. So, it is compatible with what the scale knows (together with our presuppositions about it) that it knows ϕ_{10}, while ϕ_{11} is false.

Note that this critique of the anti-luminosity argument takes a different form from Berker's, although both critiques are based on a claim of a constitutive connection between the candidate for a luminous condition and knowledge or belief that that condition obtains. Berker's argument was that the safety requirement for knowledge did not imply the margin-of-error premises, while my argument involves a rejection of Williamson's safety requirement itself. In the artificial example, the device "knows" a proposition even though in very similar situations in which it still "believes" the proposition it is false, violating Williamson's safety condition. But while it is clear that this safety condition fails on the assumption of the schematic account, if we are to extend this critique to the case of human knowers, it must be argued that the rejection of this safety condition is intuitively plausible.

First note that it will be very unlikely that the actual value should be at the limit of the scale's range of accuracy. If the scale is *always*, when it is functioning normally and atmospheric conditions are also normal, within the range ± 1, then it will *almost* always be within a narrower range. If we were to learn that the scale registered 42 when the actual weight was within a milligram of being out of range, we would think that *probably* conditions were abnormal. But still, it is *possible* that conditions are normal when the actual weight is 41.0 and the scale registers 42. In this case, according to our model, the scale knows that the weight is 42 ± 1.

So, suppose our scale is actually in a situation like this. There are exactly 41.00 grams of coffee beans in the bowl, the scale registers 42, and the relevant normal conditions are satisfied. We now remove one small fragment of a bean from the bowl—just enough to take the actual weight down to 40.99. It could happen that the reading on the scale then drops to 41, but suppose it stays at 42. We can then conclude that the channel conditions are no longer satisfied, since there is a constitutive connection between the channel conditions and the margin of error that defines the content of the information registered, a connection that guarantees that *if* the conditions are satisfied, the actual weight is within the range. The "belief" that the weight was 42 ± 1 was *true* before we removed the fragment of the bean, but because the belief persists and is false in the very similar situation that results from removing the small bean fragment, the belief was not *safe* in the prior situation, in the sense defined by Williamson. Should we take this as a sign that our artificial model leaves out something that is essential to a more realistic conception of knowledge? I will argue that it does not—that Williamson's safety condition does not capture an intuitive notion of safety that it is reasonable to impose on an account of knowledge.

We can grant that there is some artificiality in the precise cutoff point, but whether or not knowledge must satisfy a safety condition, there will inevitably be a sharp and somewhat arbitrary line (in an artificial sharpening of a vague and context-dependent notion of knowledge) between the situation in which the knowledge claim is true and a very similar situation in which it is false. The question is whether it helps—whether it somehow makes the knowledge safer, in a situation on the cusp of this cutoff point—to require that the belief still be true on the other side of the line when it is no longer knowledge. No matter how we understand knowledge, any case just on the other side of the line will be a case in which the agent has a belief, and presumably a justified belief, that is not knowledge. If we assume Williamson's safety condition, then the situation, just on the other side of the line, will be one in which the agent has a *true* belief that is justified, but that is not knowledge. Does it make a belief safer, in a sense of safety that has epistemic merit, if all the very similar situations that are not cases of knowledge are Gettier cases (cases of justified true belief without knowledge) rather than cases of false belief? Gettier cases (on one natural diagnosis) are cases where the fact that the proposition believed is true does not play an appropriate role in bringing about or sustaining the belief. In Gettier cases, it is just a coincidence that the justified belief is true. However the relevant nearness relation is spelled out, it does not seem reasonable to think that a belief being true by coincidence in a nearby situation should contribute to the robustness and stability of the belief in the actual situation.

The threshold cases in the sorites-like series used in Williamson's anti-luminosity arguments might not seem like Gettier cases, but in the case of the simple measuring instruments, it is, in a sense, accidental whether the "believed" proposition remains true just on the other side of the threshold point. Consider the case of the oven thermometer. If the distribution of kinetic energy within the oven is sufficiently uneven, and so improbable, then conditions are abnormal. In abnormal conditions, the temperature reading is unreliable, since the temperature in the space near the sensor might easily diverge from the average. But however we draw the cutoff line between normal and abnormal, the reading *might*, by chance, be correct, even when conditions are abnormal: it might be that, by chance, the average of the uneven temperatures in the different parts of the oven balance out to be the same (within the margin of error) as the temperature at the point of the sensor. At the point at which the thermodynamical conditions shift from being normal to abnormal, the shift could at the same time bring about a divergence between the instrument's reading and the actual average temperature, or it could leave it within range. But this should not matter for the reliability of the thermometer just before the threshold point. It is, at that point, just barely reliable, either way, and there must be a point (on any artificial sharpening of the notion of reliability) at which a reliable instrument is just barely reliable.

The arbitrariness of the cutoff point is particularly salient in this case, and it will presumably be true in cases of the knowledge of real agents that there will be some continuous gradations relevant to distinguishing what is known from what is not. With both simple artificial examples, and with realistic situations involving real human knowers, the line might be drawn in different places in different contexts. It might be when the theorist or attributor focuses her attention on threshold cases, there is some temptation to shift to stricter channel conditions, and a resulting narrower margin of error.[18] But doing so changes the interpretation of the content of what information is registered by the knower. The context change changes not only the second-order claim about what is known about what is known, but also the first order knowledge claim.

In rejecting Williamson's safety and margin-of-error principles, I am not rejecting the idea that knowledge must be safe. Knowledge, on the kind of account I am promoting, is attributed relative to background conditions that are presupposed to hold (by both the attributor and the agent) when knowledge is attributed or expressed. It is correct to say, on this kind of account, that when a person has knowledge, that knowledge is perfectly safe, *conditional on those presupposed conditions*. Under those conditions, the knowledge claims are *guaranteed* to be correct

[18] Cf. Graff-Fara 2000.

when the agent is in the relevant internal state. Of course, the presuppositions might be false, and then all bets are off. But that is the situation that we poor cognitively homeless people face, as Williamson has taught us.[19]

The background conditions must be *true*, for the knowledge claim to be true, but they need not be *known* to be true by the agent. The idea that there are conditions that are necessary for a knower to have knowledge, but that need not themselves be known, is not just a feature of our simple artificial examples of information-carrying instruments. It is a familiar claim, for example, that acquiring knowledge by perception requires that the relevant perceptual systems are functioning normally, but does not require that the person with perceptual knowledge knows that this condition is satisfied.

Still, there are crucial differences between the measuring instruments that I have used as simple models of knowers and the more sophisticated devices that we are, and the differences are relevant to the KK thesis. The simple measuring instruments have a single fixed channel for the receipt of information, and a limited range of facts that they are capable of carrying information about. My humble scale cannot ask itself whether it really knows what our external account says that it knows; it can't ask whether it is functioning normally, whether the weight that it registers is the actual weight of the beans. While it is not required, with either simple measuring devices or real knowers, that the knower knows that the background conditions obtain, it is still true that reflective human knowers, unlike simple devices, can raise the question whether these conditions do obtain. That is, a human knower can consider whether she has reason to believe the propositions she was presupposing when she represented herself as knowing something. One can ask, for example, whether one's perceptual systems are functioning properly, whether one is reasoning correctly in assessing the evidence, whether what one takes to be evidence is evidence that one really has.

The fact that we, unlike the simple information-registering devices, can ask these questions does not imply that that we are not information-carrying systems to which a version of the schematic model applies. A human knower, unlike my scale, acquires its knowledge through multiple channels, and has devices for integrating the information received from different sources, and for assessing the reliability of some of its sources with the help of others. This is what makes it possible for it to raise questions about its own reliability, and to go some way toward answering them. Even the scale takes a small step in this direction: it sends a flashing signal when the battery is low—in effect, registering information about the reliability of its basic measurement. So, the fact that the low-battery signal is *not* received carries the information (when *its* conditions are normal) that the conditions for the basic measurement are, in one respect, normal. But, of course, the low-battery indicator might malfunction as well. One cannot draw too much comfort from the fact that a witness assures you that he is a reliable witness, or that your tests of your visual system assure you that it is working well.

[19] "We must... accept the consequences of our unfortunate epistemic situation with what composure we can find. Life is hard" (Williamson 2000, 237).

The complexity that allows an information-registering device to register information about its own reliability is relevant to the KK thesis since it is natural to take the question whether one knows that one knows some fact as a question of this kind. But to raise a question of this kind—to question the presupposed background conditions relative to which a knowledge claim was made—is to change the context, and when the context is changed in this way, it brings into question the original knowledge claim. Consider any intuitively natural situation in which an agent raises the question, about some proposition he was representing himself as knowing, "Do I really know that?". If on reflection the answer is, "I'm not sure," or "probably not," then the agent will retract the original claim. It is not just that he will refrain from asserting that he knows it; he will refrain from making the first-order assertion. The argument I gave for the KK principle, in the context of the schematic model of a device that functions to carry information, assumed that the context remained fixed in the following sense: the KK proposition was assessed relative to the same presupposed channel conditions as the first order K proposition. With real knowers there is much more scope for contextual variation, many more dimensions on which the context for a knowledge claim may vary, but I think it is reasonable to assume that the argument for KK should work, on this assumption, even when we are talking about much more sophisticated measuring devices such as ourselves.

So my defense of the KK principle can allow for cases where the presupposed channel conditions for a first order knowledge claim are different from those presupposed for a second order claim about what one knows, and so can allow that the first might be true, in its context, while the second is false in its different context.[20] But the conclusion that the KK principle holds on the assumption that the background presuppositions are held fixed is still a significant thesis, and is a thesis that is incompatible with the premises of Williamson's anti-luminosity argument. On the conception of knowledge that Williamson is defending—one that satisfies his safety and margin-of-error principles—the claim that one knows that one knows that ϕ requires stronger evidence *about* ϕ (and not just about the conditions necessary for knowing ϕ). On this view, a reflective and self-aware person might rationally judge that her evidence is sufficiently strong to assert that ϕ, but not sufficiently strong to assert that she knows that ϕ. So the assertion, "ϕ, but I am not sure that I know that ϕ," or even, "ϕ, but it is highly probable that I don't know that ϕ," might be appropriate assertions that are in conformity with a knowledge norm of assertion (the norm, defended by Williamson, that is violated by an assertion of a proposition that is not known by the speaker). But one might think that a reflective and self-aware knower could reason this way, when she finds herself prepared to assert that ϕ: "I *could* be in a situation where ϕ is false, or one in which it is true, but I don't know it. (I'm not infallible. It has on occasion happened that I took myself to be in a position to make an assertion, but it turned out that I was mistaken.) But while I recognize that this theoretical possibility is always there, I am not a skeptic, and I don't take this theoretical possibility to imply that not-ϕ is an *epistemic* possibility for me, in my

[20] Dan Greco, in Greco 2014, emphasizes the context shift involved in apparent counterexamples to the KK principle.

present situation. My considered judgment is that I am now in an evidential position that justifies as assertion that ϕ. Am I also in a position to assert that I know that ϕ? Well, in a sense, I would be *as* safe in asserting that I know that ϕ as I am in asserting that ϕ, since if I don't know that ϕ, then the first-order assertion by itself is in violation of the norm. But if my first order assertion is in conformity with the norm, then the second-order claim about my knowledge is *true*. Why isn't this enough to make that second-order assertion a *safe* assertion, and why isn't it enough to justify me in claiming that if I indeed do know that ϕ, then I know that I do?" While my defense of the KK principle rested on an analogy with simple artificial information-registering devices, and a highly contentious theoretical hypothesis that a schematic model for such devices generalizes to real human knowers, I am also inclined to think on more intuitive grounds that this reflective, self-aware (if perhaps slightly neurotic) knower is making perfectly good sense.[21]

[21] Thanks to Dan Greco and Damien Rochford for helpful discussion of these issues.

3

Iterated Belief Revision

1. Introduction

There is a relatively well-established formal theory of belief revision—the so-called AGM theory, named for Carlos Alchourrón, Peter Gärdenfors, and David Makinson, who first proposed it in the 1980s. Agents' dispositions to revise their beliefs are represented, in this theory, by functions taking prior belief states into posterior belief states that are determined by an input proposition, the information that induces the revision. The theory states constraints on such a function, and the constraints that were proposed gave rise to a nice model-theoretic structure. The initial theory did not give any account of iterated belief revision, but it has been extended in various different ways, with constraints on the way that a posterior belief state might be further revised by a sequence of input propositions. The iterated revision theories are more varied, and more controversial. The constraints proposed by different theorists are sometimes in conflict with one another, and they are (I will argue) less well motivated than at least the basic AGM postulates. So, despite the existence of a number of very elegant proposals for solving the problem of iterated belief revision, I think the problem has not been solved. I will argue that we need to go back to the beginning and look more closely at the nature of the problem, and at the foundational assumptions about the phenomena that an iterated belief revision theory is trying to clarify. And I will argue that little of substance can be said about iterated belief revision if we remain at the level of abstraction at which such theorizing has been carried out. We need to distinguish different kinds of information, including meta-information about the agent's own beliefs and policies, and about the sources of his information and misinformation.

So, I will start by going back to the beginning, setting up the problems of belief revision, and iterated belief revision. While my ultimate concern is with foundational issues, I will spend some time trying to get clear about the shape of the formal problem, and of the abstract structure of some of the solutions to it that have been proposed. I will then look critically at some of the postulates that have been proposed, at the general considerations that motivate them, and at some examples that raise problems for them. I will be concerned both with the assessment of the particular postulates, and with more general questions about the dialectic of critique and defense in such cases.

I will discuss the basic problems, and the various solutions, entirely in a model theoretic framework. The original AGM theory was given a syntactic formulation, and much of the subsequent discussion has followed this lead. Belief states were represented by sets of sentences, and belief changes were induced by sentences. But it was assumed from the beginning that the relevant sets were deductively closed, and that logically equivalent sentences played an equivalent role, so a model theoretic framework that represents belief states by sets of possible worlds, and information more generally by sets of possible worlds, can capture the same essential structure, and it was recognized early on that the AGM theory could be given a simple model-theoretic representation.[1] For several reasons, I think the model theoretic representation has conceptual as well as technical advantages, avoiding some potential for confusion that may arise from the syntactic formulation. First, the status of the language in terms of which information is represented in the syntactic formulation is unclear. It is not the language used by the subjects whose beliefs are being modeled (they do not generally speak in propositional calculus), though it is often treated as if it is. Second, the syntactic formulation brings a potential for use-mention confusion, and confusion of properties of the sentences with properties of what those sentences are being used to say.[2] Third, there are additional problems when the sentences in question are context-sensitive, expressing different propositions at different times or in different epistemic contexts. Conditional sentences, which are sometimes used to represent belief revision policies, and tensed sentences about what one believes at the present time are both obviously context-sensitive in these ways, and there is a potential for confusing the identity of a sentence, across contexts, with the identity of what it is being used to say.[3] Finally, a further reason to prefer the model-theoretic representation is that our concern is with knowledge, belief, and belief change, and not with speech, and we do not want to make implicit assumptions about the forms in which information is represented, or about the linguistic capacities of the subjects we are modeling. There may be rational agents who act, but do not speak, and more generally, there may be contents of knowledge and belief that the agent lacks the capacity to articulate. A general theory of belief revision should apply in these cases as well as to the case where what an agent knows corresponds to what he is in a position to say.

So, in my formulation of the various theories, there will be no languages involved. Of course, we use language to formulate our problems and theories, but so do geologists when they are talking about rocks. As with geologists, languages will form no part of the subject matter of our theory. The primitive elements of our basic structure are possible worlds, and the propositions that are the items of information that may be believed by an agent in a prior or posterior belief state will be sets of possible worlds. Propositions are thus identified with their truth conditions—with the distinction between the possibilities in which they are true and the possibilities in which they are false.[4]

[1] See Grove 1988.

[2] I discuss some of the potential confusions, in the context of a discussion of theories of non-monotonic reasoning, in Stalnaker 1994.

[3] I think some of the issues about updating vs revision would be clearer if sentences were more clearly distinguished from what they are used to say.

[4] Any representation of cognitive states that uses this coarse-grained notion of proposition will be highly idealized. It will be assumed that agents know or believe all the logical consequences of their knowledge or beliefs. This is a familiar feature of all theories in this general ballpark, including probabilistic representations of degrees of belief. The syntactic formulations make an analogous idealization. There are

One way to develop, in this model-theoretic context, a representation of knowledge and belief is to characterize propositions *about* the agent's knowledge and belief, as in Jaakko Hintikka's logics of knowledge and belief. This approach usually takes the form of a semantics for a language with knowledge and belief operators, but the language is dispensable: one can enrich one's model (the initial model being just a set of possible worlds) by giving it the structure necessary to define a function taking any proposition α (represented by a set of possible worlds) to the proposition that the agent knows α. The required structure is a Kripke structure, or frame: a set of possible worlds and a binary relation R on it, where "xRy" says that possible world y is compatible with what the agent knows or believes in world x. But the AGM belief revision theory did not follow this path. In its model-theoretic version, a belief state is represented simply by a set of possible worlds—those compatible with what is believed. The theory is more abstract than the Hintikka-style theory in that there is no explicit representation of knowledge and belief about what the agent knows or believes. This is a significant difference, but it is important to recognize that this theoretical decision is not a decision to ignore information about the agent's knowledge and belief, or to restrict in any way the kind of information that might be relevant to inducing a belief change, or more generally in distinguishing between the possible worlds. Rather, it is a decision to theorize at a level of abstraction at which nothing is said, one way or another, about the subject matter of the information. Meta-information about the agent's knowledge and belief is like information about politics. The theory lacks the resources to distinguish information about politics from information, for example, about astronomy, but of course information about politics or astronomy might be information that is compatible with an agent's beliefs, or that induces a change in belief. Similarly, the abstract AGM theory lacks the resources to distinguish meta-information from information about the physical environment, but nothing forecloses the possibility that information about informational states be relevant. We cannot generalize about the distinctive role of such information, if it has a distinctive role, but when we consider examples where we describe a certain concrete scenario, and then construct a model to fit it, if the scenario is one in which meta-information is intuitively relevant, then we must include it in our model. (A corresponding point might be made about the syntactic formulation. The primitive sentence letters, in such a formulation, have no special status: they represent the general case. Any generalization that one makes should continue to apply if one enriches the language to include quantifiers, modal operators, epistemic operators, causal conditionals, or whatever.)

2. The AGM Belief Revision Theory

The basic belief revision problem is to model the result of the process of changing one's state of belief in response to the reception of a new piece of information. Assume that we have a set of possible worlds, B, representing a prior belief state, and

various different ways of motivating the idealization, and of explaining the relation between theory and reality on this issue, but that is a problem for another occasion. Most will agree that the idealization has proved useful, despite its unrealistic character.

a proposition α representing the new information. A belief revision function will be a function $B(\alpha)$, whose value will be the posterior belief state that would be induced by this new information. The problem is to specify the constraints on such a function that any rational belief revision must satisfy, and to motivate those constraints. Let me sketch the AGM theory's solution to this problem, as spelled out in the model-theoretic context, and then turn to the further problem of iterated belief revision.

To model a belief revision function, we start by specifying two sets, B and B*, the first a subset of the second. B represents the prior belief state, and B* is the set of possible worlds that are candidates to be compatible with some posterior belief state. An AGM belief revision function will be any function from propositions (subsets of B*) to posterior belief states (also subsets of B*) that meets the following four conditions for all propositions:[5] (I will use 'Λ' for the empty set of possible worlds.)

AGM1 $B(\alpha) \subseteq \alpha$

AGM2 If $B \cap \alpha \neq \Lambda$, then $B(\alpha) = B \cap \alpha$

AGM3 If $B^* \cap \alpha \neq \Lambda$, then $B(\alpha) \neq \Lambda$

AGM4 If $B(\alpha) \cap \beta \neq \Lambda$, then $B(\alpha \cap \beta) = B(\alpha) \cap \beta$

AGM4, as we will see, is less obvious and more problematic than the other postulates, so one might consider a more cautious theory, which I will call AGM– defined just by the first three postulates. One might also consider intermediate theories that are stronger than AGM–, but weaker than the full AGM.[6]

Any AGM function gives rise to a nice formal structure. It is equivalent to a formulation in terms of a binary ordering relation, or in terms of an ordinal ranking function on the possible worlds: worlds compatible with B get rank 0, the most plausible alternatives get 1, the next get 2, etc, with all worlds in B* getting some ordinal rank. The ranking of worlds determines a ranking of propositions: the rank of proposition α is the minimal rank of possible worlds within α. Any ranking function of this kind will determine an AGM revision function, and any AGM function will determine such a ranking function. In the ranking-function formulation, $B(\alpha)$ will be defined as$\{w \in \alpha: r(w) \leq r(\alpha)\}$.[7]

[5] The usual formulation of the AGM theory, in the syntactic context, has eight postulates. One of them (that logically equivalent input sentences have the same output) is unnecessary in the model-theoretic context, since logically equivalent propositions are identical. Analogues of my AGM1 and AGM4 are, in the usual formulation, each separated into two separate conditions. Finally, the first of the usual postulates is the requirement that the output of the revision function is a belief set, which is defined as a deductively closed set of sentences. This is also unnecessary in the model theoretic context. In the syntactic formulation, there is no analogue of our B*: it is assumed that every consistent and deductively closed set of sentences is an admissible belief set.

[6] Hans Rott discusses such theories in Rott 2001.

[7] It should be emphasized that ranking functions of the kind I have defined are not the same as Wolfgang Spohn's ranking functions. Or more precisely, they are a special case of Spohn ranking functions. Spohn's ranking functions are richer structures than AGM ranking functions, since they allow the values of the function to be any ascending sequence of non-negative integers. In a Spohn function, it might happen that there are gaps in the ranking (a positive integer k such that some worlds have ranks greater than k, and some less, but none with rank k), and the gaps are given representational significance when the theory is extended to an iterated revision theory. But any Spohn ranking function determines a unique AGM structure, and any AGM structure determines a unique Spohn ranking function with no gaps. See Spohn 1988.

Why should these four postulates constrain any rational revision, no matter what the subject matter? To begin answering this, we need to say more about exactly what we are modeling. First, exactly what does it mean to believe, or accept a proposition? Second, what is the status of the argument of the function—the proposition that induces the change?

On the first question: I will assume that we are talking about a strong, unqualified doxastic state, although the postulates may be appropriate for various weaker notions of acceptance as well. Specifically, I will assume that to believe α is to take oneself to know it.[8] Belief in this sense differs from knowledge only in that believing α is compatible with α being in fact false, or a Gettier case of justified true belief. And, of course, believing α is (at least in some cases) compatible with later discovering that it is false, and so being forced to revise one's beliefs so that one believes the complement of α. Actual knowledge is in a sense unrevisable, since one cannot discover (come to know) that what one knows is false. But one may know something, while still having rational dispositions to respond to potential information that conflicted with one's knowledge. Suppose I know who won the gold medal in a certain Olympic event—Michael Phelps. Still, my system of beliefs would not collapse into contradiction if, contrary to fact, I discovered that I was mistaken. In such a counterfactual situation, I would revise my beliefs to accommodate the belief-contravening evidence, with exactly how I revise depending on the details of that evidence. Of course, my rational dispositions to revise in response to contrary evidence won't in fact be exercised, since he did win the gold in the event in question, and I know that he did.[9]

On the second question: I will assume that the input proposition represents an item of information that the subject takes himself to have come to know (in a situation in which the disposition to revise is exercised), and that it is the total relevant information received. That is, the rational disposition being modeled is the disposition to shift to the posterior belief state upon coming to know (or taking oneself to have come to know) the input proposition, and no stronger new piece of information. The total evidence assumption is, of course, essential to the application of the theory, since any nontrivial belief revision function that permits revision by information that is incompatible with prior beliefs must be nonmonotonic. One may come to believe something on receiving information α that one would not come to believe on receiving information that entails α.

[8] By "taking oneself to know" I do not intend a reflective state of believing that one knows, but just a cognitive state that is like knowledge in its consequences for action. I also think that in the idealized context of belief revision theory, it is appropriate to make the kind of transparency assumptions relative to which taking oneself to know, in the relevant sense, entails believing that one knows, but that is a controversial question that I need not commit myself to here.

[9] So I am assuming a different conception of knowledge from that assumed by, for example, in Friedman and Halpern 1999. They assume that to take observations to be *knowledge* is to take them to be unrevisable. In terms of our notation, they are, in effect, identifying knowledge with what is true in all possible worlds in B*. Friedman and Halpern's way of understanding knowledge is common in the computer science literature, but I think it is a concept of knowledge that distorts epistemological issues. See Stalnaker 2006 for a discussion of some of the questions about logics of knowledge and belief.

Now let's look at the four AGM postulates in the light of these answers to our two questions. AGM1 is unassailable, given our answer to the second question. To question it is to reject the intended application of the theory. AGM2 I take to be implied by the answer to the first question, together with a proper account of what it is to know or fully believe a proposition. The crucial assumption that I want to make here is that to fully accept something (to treat it as knowledge) is to accord it this privileged status: to continue accepting it unless evidence forces one to give up something. This is a relatively weak commitment: it says only that if one accepts α, then one will continue accepting α provided the evidence one receives is compatible with *everything* that one accepts. But it is a commitment that distinguishes full acceptance from any degree of belief, no matter how high, that is short of probability one.

AGM3 also seems unassailable, given the answers to the second question, and given our assumption about what the set B* represents. B*, by definition, contains all the possible situations that are compatible with any proposition that one might conceivably come to accept, and coming to accept such a proposition means coming to be in a coherent belief state.

The intention, in the attempts to justify the first three AGM postulates, has been to show that the postulates are constitutive of the concepts involved in the application of the theory. The claim is that these principles do not give substantive guidance to inductive reasoning, but are implicit in the relevant concept of acceptance, and the intended application of the theory. But I don't see how to give AGM4 the kind of justification that we can give for the other postulates. (One temptation is to justify this postulate in virtue of the nice formal structure that it determines, but this is a different kind of justification, and I think one that we should resist.) I will defer discussion of the motivation for AGM4 until we consider the iteration problem, since I think that if it can be justified at all, it will be in that context. So, let me now turn to that problem.

3. The Iteration Problem

We started with simple belief states, but the belief state itself, as represented formally, was not a rich enough structure to determine how it should evolve in response to new evidence. So, we added new structure, which yielded a posterior belief state for any given input proposition. One might put the point by saying that a full representation of an agent's cognitive situation should be represented, not by a simple belief state, but by a belief revision function. But the output of an AGM function is just a simple belief state, and so it does not yield a full representation of the agent's posterior cognitive situation. We need to enrich the theory further so that the output of our revision function is a new revision function. But, of course, if we enrich the representation of the cognitive situation further, then we need to be sure that we also get a structure of the enriched kind as our output. We need to satisfy a principle that Gärdenfors and Rott have called "the principle of categorical matching"[10]: roughly, that our representation of a cognitive situation must be a kind of structure such that given a proposition as input, yields a structure of the same kind as output.

[10] Gärdenfors and Rott 1995, 37.

Before looking at some particular proposals for solving this problem, let me describe a general form that any solution can take, and introduce some terminology for talking about the solutions.[11] Any solution to the iteration problem will yield a function taking finite sequences of propositions to belief states.[12] Call any such function a *belief system*. The idea is that the proposition that is the first term of the sequence induces a certain revision, and then the next term induces a revision of the belief state that results from the first revision, etc. The initial prior belief state of a belief system will be the value of the function for the empty sequence. A belief system, restricted to inputs that are one-term sequences will be a revision function (a function from propositions to posterior belief states). Say that a revision function R is *generated* by a belief system Ψ iff for some sequence of propositions, $\beta_1, \ldots \beta_n$, and for all propositions α, $R(\alpha) = \Psi(\beta_1, \ldots \beta_n, \alpha)$. Say that a belief system Ψ is an AGM belief system iff every revision function that it generates is an AGM function, and similarly for AGM– belief systems.[13]

The weakest and most unconstrained "solution" to the iteration problem would be simply to note that an AGM belief system yields a belief state and an AGM revision function for any input. (And any AGM– belief system yields an AGM– revision function for any input.), so we can just say that an agent's full cognitive situation is to be represented by such a belief system. But one would like to find some more interesting and revealing structure that determines such a system, and some constraints on the way that belief states may evolve.

The most constrained kind of solution would be to lay down postulates governing belief systems that are strong enough so that there will be a unique full belief system determined by any initial AGM belief revision function. Craig Boutilier proposed a theory that meets this condition in Boutilier 1996. A more flexible approach would be to state and defend a set of postulates that constrains the evolution of a belief system, but that allows for alternative evolutions from the same initial revision function. I will describe Boutilier's system, and a classic proposal of the more flexible kind (by A. Darwiche and J. Pearl) before looking at the ways that such proposals are motivated and evaluated.

Boutilier's proposal states a rule for taking any prior AGM revision function and any (consistent) input proposition to a new AGM function. The idea can be put simply, in terms of the ranking-function representation of an AGM revision function. Let r(w) be the rank of possible world w according to the prior AGM ranking

11 The general form I will describe was used by Lehmann 1995. A related formalism is used in Rott 1999.

12 Recall that a belief state consists of a *pair* of sets of possible worlds, B and B*, the first being a subset of the second. In the general case of a belief system, it will be important to consider the case where B* as well as B may take different values for different arguments of the belief system. But to simplify the discussion, I will assume for now that for any given belief system there is a fixed B* for that system.

13 As we defined AGM revision functions, the input proposition could be an impossible proposition (the empty set). In this vacuous limiting case, the postulates imply that the output is also the empty set. This was harmless in the simple theory. The empty set is not really a belief state, but it doesn't hurt to call it one for technical purposes. But we need to do some minor cleaning up when we turn to the extension to a full belief system. One simple stipulation would be to require that in the empty "belief state," the B* is also empty. Alternatively, one might restrict the input sequences to sequences of nonempty propositions, or in the general case where B* may change with changes in the belief state, to sequences of nonempty subsets of the relevant B*.

function, and let $r_\alpha(w)$ be the rank of w in the posterior AGM ranking function induced by information α. The Boutilier rule (with one minor simplification) is this: $r_\alpha(w) = 0$ for all $w \in B(\alpha)$, and $r_\alpha(w) = r(w) + 1$ for all $w \notin B(\alpha)$.[14] Intuitively, the idea is to move the posterior state into the center of the nested spheres, but to leave the ordering of all the other worlds the same.

How might such a solution be justified? It is, in one clear sense, the uniquely minimal revision of an AGM revision function, and so it might be thought to be a natural extension of the minimality assumption that is implicit in postulate AGM2. But we suggested that the justification of AGM2 was not some general methodological principle of minimality, but rather an assumption about the kind of commitment that is constitutive of accepting, or fully believing a proposition. According to this way of understanding it, AGM2 does not in any way constrain a rational agent's response to an event that one might be tempted to describe as the receipt of a certain piece of information. On the intended application of the belief revision theory, an event is *correctly* described as the receipt of the information that α only if the event is one in which the agent fully accepts α, and so undertakes the commitment. An event that for one agent, or in one context, constitutes receiving the information that α might not be correctly described in that way for another agent, or in another context. One might, for example, respond to an observation that it seems to be that α, or a report by a witness that α, with a commitment short of full belief, perhaps concluding that α has a very high probability, but less than one. AGM2 does not say that one should or should not respond in this way: it says only that *if* one responds by fully accepting the proposition, then this constitutes taking on a commitment to continue accepting it until one is forced to give something up.

This kind of justification for a postulate, as a constraint on any rational revision, does not extend to any kind of minimality assumption in the iterated case. It is not constitutive of having the epistemic priorities represented by a belief revision function that one is disposed to retain those priorities in response to surprising new information.

But one might be satisfied, in one's search for a solution to the iteration problem, with a more relaxed kind of justification. A demanding justification of the kind that I suggested could be given for the postulates of AGM–, requires an argument that given what it means to fully believe something, and given the assumption that the revision function applies only when the input proposition is fully believed, it would be irrational to violate the postulates. Or perhaps it would be better to say that a successful justification of this kind would show that if you apparently violated the postulate in question, that would show that the conditions for the application of the theory were not met. (You didn't really fully accept either the input proposition or one of the propositions that characterize the prior belief state.) But even if one cannot give a justification that meets this demanding standard for the postulates of an iterated belief revision theory, one might argue that a weaker kind of justification

[14] A pedantic qualification is needed to get this exactly right. In the case where B(α) is the set of *all* possible worlds of a certain rank greater than 0, the Boutilier rule, as stated, will result in a ranking function with a gap. If a simple AGM function is represented by a ranking function with no gaps, one needs to add that after applying the rule, gaps should be closed.

would suffice. According to this line of thought, there may be very general and abstract substantive methodological principles that govern the appropriate formation and change of belief, and the project of developing belief revision theory is to find principles of this kind that seem to be descriptively adequate, and that seem to illuminate common-sense reasoning and scientific practice. If one thinks of the project this way, then one should assess a proposed set of postulates by looking at its consequences for examples. If the consequences seem intuitively right, that is reason enough to accept them. This is a reasonable aim, though I think it is important to distinguish the more stringent kind of justification from the more relaxed kind.

Before considering this kind of evaluation of solutions to the iteration problem, I will sketch an influential proposal of the more flexible kind: a set of constraints that permits alternative evolutions of a given initial AGM revision function. The iteration postulates proposed in Darwiche & Pearl 1997 give rise to a nice formal structure; they allow Boutilier's theory as a special case, but avoid some of the problems that have been raised for that more constrained proposal.

Darwiche and Pearl propose four postulates to extend the AGM theory. I will state them in terms of the framework and terminology introduced above. A DP belief system Ψ is an AGM belief system that also meets the following four conditions for any proposition $\beta_1, \ldots \beta_n$, α and ϕ:

(C1) If $\alpha \subseteq \phi$, then $\Psi(\beta_1 \ldots \beta_n, \phi, \alpha) = \Psi(\beta_1, \ldots \beta_n, \alpha)$
(C2) If $\alpha \subseteq \neg\phi$, then $\Psi(\beta_1 \ldots \beta_n, \phi, \alpha) = \Psi(\beta_1, \ldots \beta_n, \alpha)$
(C3) If $\Psi(\beta_1, \ldots \beta_n, \alpha) \subseteq \phi$, then $\Psi(\beta_1, \ldots \beta_n, \phi, \alpha) \subseteq \phi$.
(C4) If $\Psi(\beta_1, \ldots \beta_n, \alpha) \nsubseteq \neg\phi$, then $\Psi(\beta_1, \ldots \beta_n, \phi, \alpha) \nsubseteq \neg\phi$.

Think of an AGM belief system as a procedure for redefining the ranks of the possible worlds, at each stage of the iterated process, in response to the input proposition.[15] If r(w) is the prior rank of world w, and $r_\alpha(w)$ is the posterior ranking induced by proposition α, then (C1) is equivalent to assuming that for all w,w' $\in \alpha$, r(w) > r(w') iff $r_\alpha(w) > r_\alpha(w')$. (C2) is equivalent to assuming that for all w,w' $\notin \alpha$, r(w) > r(w') iff $r_\alpha(w) > r_\alpha(w')$. (C3) is equivalent to assuming that for all w $\in \alpha$ and w' $\notin \alpha$, if r(w) < r(w'), then $r_\alpha(w) < r_\alpha(w')$. (C4) is equivalent to assuming that for all w $\in \alpha$ and w' $\notin \alpha$, if r(w) $\leq$ r(w'), then $r_\alpha(w) \leq r_\alpha(w')$. The idea is that in any belief revision induced by proposition α, all the α-worlds move down or stay the same in the ranking (with the minimal α-worlds taking posterior rank 0, and the relative order, within α, remaining the same); the non-α-worlds all move up in the ranking or stay the same (with no non-α-worlds remaining at rank 0, and the relative ranking within $\neg\alpha$ remaining the same). Given these rules, the initial AGM ranking constrains the new ranking, but does not determine it, since it does not in all cases determine the relative ranking of α and non-α worlds.

Spohn's ranking function theory provides an iterated revision rule that conforms to the DP conditions, but that is more constrained. Spohn's iterated revision rule is a reductionist account in the sense that the posterior ranking function is fully determined by the prior ranking function, plus the input. But as mentioned in footnote 7, a Spohn

[15] The ranking function representation is just a notational convenience. The theory, and all the DP constraints, could be stated without using numbers.

ranking function is richer than an AGM ranking function, since gaps in the rankings are allowed, and the gaps are intended to provide information about the entrenchment of beliefs that goes beyond ordering information. In addition, in the Spohn theory, the inputs to the revision function contain more information than is provided just by specifying the input proposition. The input will be a pair consisting of a proposition and a positive integer, where the number is intended as a measure of the degree of robustness of the new information. Any new information will be treated as something fully believed, in the posterior state, but higher levels of robustness will mean that the new information becomes more fully entrenched in the posterior belief revision structure. The Spohn rule is this (using "r" for the prior ranking function, and $r_{<\alpha,k>}$ for the posterior ranking induced by new information α, with robustness measure k):

For all $w \in \alpha$, $r_{<\alpha,k>}(w) = r(w) - r(\alpha)$.
For all $w \notin \alpha$, $r_{<\alpha,k>}(w) = r(w) + k$.[16]

Before moving on to look at concrete examples, let me mention one technical consequence of postulate (C1): it entails the fourth postulate of the simple AGM theory, the postulate that we said was more difficult to defend than the other three AGM postulates. More precisely, the result is that any AGM– belief system that satisfies postulate (C1) is an AGM belief system.[17] So if the iterated revision postulate (C1) can be defended, that will suffice for a defense of AGM4.

These are beautiful theories, providing nice formal structures, and the commentary on them provides intuitive content to the functions and relations that are specified in the structures. But are the constraints that give rise to these nice structures defensible? I will consider some examples that seem to raise problems for several of the postulates, but the task of confronting an abstract theory with concrete examples is a subtle and difficult one, and it is often questionable whether a putative counterexample is really relevant to the abstract principle to which it is applied. So before looking at the details, I will digress to make some general comments about this methodological problem.

4. Evaluating Counterexamples

Let me start with a caricature of a fallacious use of an example to refute a principle: I see a man carrying a sign reading "THE WORLD WILL END TOMORROW" (the

[16] This rule holds on the assumption that α is a proposition that is not believed in the prior state. The revision rule becomes a contraction rule when the value of k is 0.

[17] The proof is simple: By AGM2 (and the assumption that Ψ is an AGM- system),

1. If $\Psi(\beta_1,\ldots\beta_n,\phi)\cap\alpha\cap\phi \neq \Lambda$, $\Psi(\beta_1,\ldots\beta_n,\phi,\alpha\cap\phi) = \Psi(\beta_1,\ldots\beta_n,\phi)\cap\alpha\cap\phi$

But by AGM1, $\Psi(\beta_1,\ldots\beta_n,\phi)\cap\alpha\cap\phi = \Psi(\beta_1,\ldots\beta_n,\phi)\cap\alpha$, so it follows from 1 that

2. If $\Psi(\beta_1,\ldots\beta_n,\phi)\cap\alpha \neq \Lambda$, $\Psi(\beta_1,\ldots\beta_n,\phi,\alpha\cap\phi) = \Psi(\beta_1,\ldots\beta_n,\phi)\cap\alpha$

By (C1),

3. $\Psi(\beta_1,\ldots\beta_n,\phi,\alpha\cap\phi) = \Psi(\beta_1,\ldots\beta_n,\alpha\cap\phi)$

Therefore, by 2 and 3,

4. If $\Psi(\beta_1,\ldots\beta_n,\phi)\cap\alpha \neq \Lambda$, $\Psi(\beta_1,\ldots\beta_n,\alpha\cap\phi) = \Psi(\beta_1,\ldots\beta_n,\phi)\cap\alpha$

But 4 is just the claim that AGM4 holds in all revision functions generated by Ψ.

kind of event that takes place mostly in *New Yorker* cartoons). So, I receive the information that the world will end tomorrow, but I don't revise my beliefs by coming to fully believe that the world will end tomorrow. I think the man is a nut, so my beliefs about the future of the world are unaffected by the message on his sign. Therefore, AGM1 (which says that the input proposition is fully believed in the posterior belief state) is not in general true.

Mathias Hild, in Hild (1998) discussed examples something like this, and he seemed to be using them as counterexamples to a principle analogous to AGM1. (Hild's first example was of testimony from a person that the subject believes is trying to be misleading, and so she not only fails to believe the statement made, but takes the fact that it was made as decisive evidence that it is false.) I am not sure what Hild had in mind with these examples, but he seemed to be suggesting that we should characterize the evidence, observation, or information that induces a belief change independently of its effect on the cognitive situation of the subject whose beliefs are changed. But I am not sure how to do this in general, and I don't think it is a good idea in any case, to try. It is not the job of belief revision theory to say what a person should come to believe on having certain kinds of experiences. The task is to say how new beliefs interact with a prior belief state to determine a posterior state.

Hild says "receiving total evidence *B*, as I understand it, means to be presented (precisely) with the information that *B* has occurred, or to receive (precisely) the input that *B*."[18] But this is subject to different interpretations. Perhaps in a sense, when someone tells me (in speech, or on a sign) that P, I am presented with the information (or misinformation) that P. The statement is surely *evidence*, but a statement that P is not (necessarily) evidence that P—it is evidence that someone said that P.

What is meant by '*evidence*'? Perhaps in one sense, evidence may consist of things such as hair samples, pieces of a broken vase, a gun with fingerprints on it, that one might label and put in plastic bags until the trial. Such things *carry* information, and that is why they might be entered into evidence, but they are not themselves items of information, and that is what an input to a belief revision function must be. Timothy Williamson argues persuasively (in Williamson (2000)) that evidence should be identified with knowledge. An input proposition need not become known, in the posterior belief state that it induces (and so need not be evidence, according to Williamson's thesis), since rational agents may make mistakes. But we are assuming that any input proposition is (in the induced posterior belief state) taken by the subject to be knowledge, or at least that it becomes fully believed. This is not a claim made by the theory, but a constraint on its application.

But there are more subtle cases. Suppose I am disposed to take the word of either of two potential informants, Alice and Bert (at least on certain subjects). If either Alice or Bert were to tell me that the Red Sox won, I would take myself to have learned that they did. Both Alice and Bert are a lot better informed than I am about current events, sports, etc., so even in some cases where I have a prior opinion (say I believe that the US won more medals than China in the 2008 Olympics), I am disposed to

[18] Hild 1998, 334.

change that belief if Alice or Bert were to tell me that I was wrong. Against this background, suppose that Alice tells me that the Red Sox won the game, so I revise by coming to believe that they did. But a little later, Bert comes and tells me that they lost, so I revise again, coming to believe that they lost. As Darwiche and Pearl say, "when two contradictory pieces of evidence arrive, the last one prevails."[19] But why should this be? It was only an accident that Alice arrived first. If she had been slightly delayed, so that I heard first from Bert, then my settled opinion, in the end, would have been the opposite. Surely it is unreasonable to require that any rational agent have revision policies that are constrained in this way. (Though the example is a case of iterated belief revision, the assumption being challenged is, again, AGM1.)

But this argument is also fallacious. The AGM principles do not say that we must give the last witness priority over earlier ones: they permit the reasonable response, which, in most case fitting this description, would be to treat the witnesses symmetrically. The assumption was only that *in the prior state*, I was disposed to take Alice's word, or Bert's, about the game. The principles imply only that if, after receiving Bert's report, I am *still* disposed to take Alice's word on the outcome of the game, then I should give Alice the last word. To have this belief revision policy is to give Alice priority over Bert, which might be perfectly reasonable, in some cases. But in such a case, I would probably reach the same conclusion, in the end, even if Alice had come first. For in that case, in the intermediate belief state (after receiving Alice's report, but before receiving Bert's), I will no longer be disposed to take Bert's word for it in cases where he contradicts what I learned from Alice.

If the epistemic status of the reporters is symmetrical, as suggested in the example, then I will most likely take neither to be reliable when contradicted by the other, so my end state will be to suspend judgment. Nevertheless, it is *consistent* for me to have the policy of letting the last reporter decide my belief, whichever one comes last. Such a policy might, in some contrived circumstances, be a reasonable one. But the postulates do not require such a policy, and that is enough to defeat the counterexample. The general moral of the story is that a concrete evidential situation (being told by Alice that P) may count as receiving information that P in one context, even if the very same situation would not be correctly described in that way in another context.

For a piece of information to be the input that induces a belief change in a proper application of the theory, it must not only be a proposition that the subject comes fully to believe, it must also be the *total* evidence—the strongest proposition that is the new information received. This condition on the application of the theory to a concrete case raises a serious general problem, since in any concrete situation where one receives evidence, one receives a lot of extraneous information along with the information that is in focus. When Alice told me that the Red Sox won, she said it in English, slowly, with a slight southern accent. She was wearing blue jeans and a grey tee shirt with a picture on it of Albert Einstein sticking out his tongue. Strictly speaking, total evidence should include all this information, everything

[19] Darwiche & Pearl 1997, 11.

that I take in while receiving Alice's report. We can't be expected to include all of this information in our models of concrete situations, but it will suffice to include all of the *relevant* information that is received. The problem is that it may be controversial whether some evidence is relevant, and one might reject a putative counterexample on the ground that it left out some relevant evidence. In response, the critic might vary the example, and show that the result claimed for the example holds across a variety of cases, where the information claimed to be irrelevant varies. The example will be unconvincing if the intuitions about the concrete case seem to be relying on information that does not get included in the model.

A further problem in the application of the abstract theory to concrete cases, concerns timing. Evidence comes in continuously, and takes time to process. In some cases, one might represent a situation either as a single belief revision, made after some complex new information is taken in and appreciated, or as a sequence of belief revisions as one takes in part of the information, revises, and then takes in more of it. It may matter, for the evaluation of iteration principles, which way one represents the case, but the choice may seem arbitrary. It would be a point against a proposed iteration principle if it had different consequences for different formal representations that seem equally faithful to the concrete case in question.[20]

In light of all these considerations, let me examine in some detail[21] an example that Hans Rott gave as a counterexample to a range of AGM principles.[22] Rott's example is a case where the information in focus concerns an appointment to be made in a philosophy department. The position is a metaphysics position, and there are many candidates, but three in particular who are of interest: Andrews, Becker, and Cortez. Andrews is clearly the best metaphysician of the three, but is weak in logic. Becker is a very good metaphysician, also good in logic. Cortez is a brilliant logician, but relatively weak in metaphysics. Now we consider two possible scenarios (these are alternatives, not successive scenarios). In scenario one, our subject Paul is told by the dean, that the chosen candidate is either Andrews or Becker. Since Andrews is clearly the better metaphysician of the two, Paul concludes that the winning candidate will be Andrews. But in alternative scenario two, the Dean tells Paul that the chosen candidate is either Andrews, Becker, or Cortez. "This piece of information sets off a rather subtle line of reasoning. Knowing that Cortez is a splendid logician, but that he can hardly be called a metaphysician, Paul comes to realize that his background assumption that expertise in the field advertised is the decisive criterion for the appointment cannot be upheld. Apparently, competence in logic is regarded as a considerable asset by the selection committee."[23] But while Paul concludes that logic

[20] Thanks to Hans Rott for raising this issue in his comments on a version of this chapter. The issue deserves more discussion.

[21] Though not in as much detail as Rott provides about the example. One should consult his discussion in Rott 2004. He includes one additional scenario, since he aims to raise problems for six different principles, as well as more detail to the example. But I think my sketch of the example includes everything relevant to the issue I am raising about it.

[22] Most of the principles Rott considers are weaker than AGM4, but entailed by the full set of AGM postulates.

[23] Rott 2004, 230.

is being given more weight than he initially thought it would be given, he still thinks that Cortez will not win out in the end, given his weakness in the advertised field, so he concludes that the winning candidate will be Becker.

Now if we assume that the total relevant information received in scenarios one is (A or B), and the total relevant information received in scenario two is (A or B or C), then we have a violation of AGM4 (as well as of some weaker principles that are implied by AGM4, but not by AGM1, 2, and 3 together). A natural response to the example is to say that some relevant evidence has been left out in the model of what is learned in scenario two. Strictly speaking, what Paul learned was that the Dean told him that the winning candidate will be either Andrews, Becker, or Cortez. Given Paul's background belief that the Dean speaks the truth, he thereby learns that either Andrews, Becker, or Cortez will be the winning candidate, but since he also learns that this is what the dean said, he learns something that he does not learn (because it is not true) in scenario one. So properly modeled, the example is not a counterexample to any of the principles.

Rott considers this objection to his counterexample. Here is how he puts the objection: "When the dean says that either Andrews or Becker or Cortez will be offered the job, isn't she in some sense saying *more* than when she says that either Andrews or Becker will be the winner of the competition? Namely that it is *possible* that Cortez will be offered the position, while the latter message, at least implicitly, excludes this possibility." I don't think this is quite the right way to put the point. It is not that the dean *said* anything more than that the winning candidate was one of those three. What is important is not what was said, but what total information Paul receives in the encounter. The point is that Paul learned that the dean said this, and he did not learn that in scenario one. The issue is whether this extra information is relevant. It seems to me that it is clear from the way Rott tells the story that it is taken by Paul to be relevant, since it is part of the story that Paul drew a conclusion from the fact that the dean included the Cortez possibility in his disjunctive statement. In scenario two, Paul concluded that Cortez was being taken seriously by the selection committee, and it is clear that the basis for this conclusion was a fact about what the Dean did and did not say.

Rott goes on to say that "nothing in the story commits us to the view that either the dean or Paul actually believes that Cortez stands a chance of being offered the position."[24] This is right, but the story, as Rott tells us, does commit us to the view that Paul believes (in scenario two) that Cortez is being taken seriously by the selection committee, and this is a conclusion he reaches only from the information received in scenario two, but not in scenario one. It is also clear from the story Rott tells that this conclusion (that Cortez is being taken seriously by the selection committee) plays an essential role in Paul's further inference that Becker will be the selected candidate. So, this extra evidence is clearly relevant evidence.

Now this extra relevant evidence is meta-information—information that Paul acquires about what he learns, and how he learns it. As it is sometimes described,

[24] Ibid., 233.

it is auto-epistemic information. Rott says, "As is common in the literature on belief formation, we presuppose in the paper that our language does not include the means to express autoepistemic possibility, something like $\diamond c$ (read as 'for all I believe, *c* is possible')."[25] But the fact that the language does not include the resources to express an autoepistemic possibility operator does not mean that information about auto-epistemic possibilities is not information that must be taken into account, if it is relevant. The simple language of belief revision theory also does not have resources to express counterfactuals as a function of antecedent and consequent, or propositions about what other people know. But the primitive sentence letters (in a syntactic representation of a belief revision theory) are unrestricted. The information they are interpreted to express could be anything, including information about how I learn some of the things I learn, about the sources of my information, or about what I believe about what I believe and don't believe. If the story we tell in an example makes certain information about any of these things relevant, then it needs to be included in a proper model of the story, if it is to play the right role in the evaluation of the abstract principles of the model.

In the end, I am not really disagreeing with Rott on the correct diagnosis of the example; if we disagree, it is about exactly what the example shows about the confrontation of abstract theory with realistic phenomena. As I read Rott's conclusion, he is using the example, as I am, to bring out the importance of meta-information. In drawing the moral of his story, he says, "What Paul really learns, in scenario 2 is not just that either Andrews, Becker, or Cortez will get the job, but that the dean says so.... The fact that a piece of information comes from a certain origin or source or is transmitted by a certain medium conveys information of its own."[26] But Rott seems to take the point about meta-information to explain why the example conflicts with the theoretical principles, whereas I want to conclude that it shows why the example does *not* conflict with the theoretical principles, since I take the relevance of the meta-information to show that the conditions for applying the principles in question are not met by the example. Rott says, "Once we make logic more realistic in the sense that it captures patterns of everyday reasoning, there is no easy way of saving any of the beautiful properties that have endeared classical logic to students of the subject from Frege on." Perhaps there is no *easy* way, but I think proper attention to the relation between concrete examples and the abstract models will allow us to reconcile some of the beautiful properties with the complexity of concrete reasoning. Rott's particular example shows that we need to take account of a richer body of information than is done in a simple model, but it does not threaten the principles of AGM theory, which should continue to apply, even when one enriches the representational resources of the language for characterizing the evidential situation. But I do agree with Rott's main point in this paper, that there are tradeoffs between the simplicity and precision that can be achieved at a higher level of abstraction, and the realism that requires a more complex immersion in the messy details of a case.

25 Ibid., 233.

26 Ibid., 237.

5. Evaluating Iteration Postulates

So, examples need to be handled with care, and the defender of an abstract principle against a putative counterexample has various moves to make to explain it away. That is why neither the defense nor the refutation of abstract principles is easy. Still, we do need to confront our theories with their consequences for the phenomena. I will consider some examples that I don't think can be explained away as easily as the examples we have discussed so far.

First, let me take a quick look, in the light of our methodological worries about examples, at a case used by Darwiche and Pearl to criticize Boutilier's iterated revision rule. In this case, I think this example stands up to scrutiny.

First, we judge that a strange new animal that we observe is a bird. Second, when we come closer we see that it is red. But then an expert informs us that the animal is in fact not a bird. Should we take back our judgment that it is red (suspending judgment on that question? Boutilier's rule implies that we must, but this is unreasonable. The source of our information that the animal was red, in this case, was independent of the earlier judgment that the animal was a bird, and so should not have been affected by the overturning of this earlier judgment. If we had observed the color first, and then later reached the conclusion that the animal was a bird, then the earlier judgment would be maintained (by the Boutilier rule). But the order in which the independent pieces of evidence are received seems intuitively to be irrelevant.

Now we added some detail to the story, beyond the fact that certain pieces of information are received in a certain sequence, namely the meta-information that the judgments were based on independent evidence. But in this case, I don't think the added information compromises the counterexample. Even if it is added (at stage two, when it is observed that the animal is red) that I also come to believe that my observation of the color is in no way dependent on the assumption that the animal is a bird, the theory will still require that, when the belief that the animal is a bird is overturned, I must give up my belief about the animal's color. So, this is a robust example that succeeds in bringing out a flaw in the Boutilier rule.

I think there are similar problems with the more flexible DP rules. The clearest problem is with (C2), but let me start with an example that has been used against (C1) (and also against AGM4).[27]

As with Rott's example, discussed above, there will be two alternative scenarios. Here is the first: there are three switches wired in such a way that the light they control is on if and only if either all three switches are up, or all three switches are down. I don't know whether the light is on or off, but I get reports from three independent observers who I take to be completely reliable. Alice tells me that switch A is up, Bert tells me that switch B is down, and Carla tells me that switch C is up, so this is what I believe. But now I learn that the light is on. I could be wrong about the

[27] An example with the structure of the example I will describe was originally given, in the context of a discussion of counterfactuals, in Ginsberg 1986. Ginsberg's story was a variation on an old example used by Quine for a different purpose. Ginsberg's example is discussed in Stalnaker 1994 and in Rott 2001, 188ff.

The example raises a problem for the basic AGM theory, and not just for the iteration rule, but as I suggested above, I think the postulate AGM4 is really an iteration principle, and should stand or fall with the DP rule (C1).

wiring arrangement, but let us suppose that my belief about that is more robust than my belief about the reliability of my informants. So, I conclude that either Bert is wrong, or else both Alice and Carla are wrong. A bold believer might reason as follows: the hypothesis of two independent errors is less plausible than the hypothesis of one error, so I should conclude that all three switches are up. But it seems too strong to force this conclusion on our subject. A more cautious believer might grant that this possibility is more likely, but refrain from reaching a definite conclusion. This more cautious policy does not seem unreasonable.

But now let us add to the story that there is also a secondary light that is controlled by just two of the switches, A and B. This light is on if and only if either both of these switches are up, or both are down. So, if the main light is on, the secondary light will be on as well, but the secondary light might be on even if the main one is not. Suppose that I knew all along about the secondary light and the way it is wired, but this does not affect the story, as told so far, since I have no prior information about whether the secondary light is on or off. It also should not affect the story, so far, if we add that the information I receive, at the last stage, is that *both* lights are on.

Now scenario two: this time, after receiving the three independent reports, but before learning anything about the state of the primary light, I learn that the secondary light is on. This new information is not relevant to Carla's report about switch C, so my belief based on Carla's report (that switch C is up) should stand. (To give it up in response to this new information would be just like giving up my belief that the animal was red, in DP's example against Boutilier.) So, I conclude that either Alice is wrong, or Bert is: either A and B are down, and C up, or else all three are up, but I don't know which. After this revision, I receive the information that the primary light is also on. Since this information is compatible with my prior belief state, all I need is AGM2 to tell me that I should conclude that all three switches are up, and in any case this seems reasonable.

In scenario two, I receive the same information as in scenario one, but in two stages: first I receive a part of the information, and then all of it. (C1) requires that I reach the same final belief state in these two scenarios, but the example suggests that in the two-stage scenario the subject should reach a more decisive conclusion. But does this example stand up to scrutiny? I find it genuinely puzzling, since I don't see that any intuitively relevant information is being excluded or misdescribed in the model. The example seems robust, in that it is easy to vary the details of the concrete story in a way that preserves the basic structure, and the variations seem equally persuasive. So, the example is not easily dismissed, but on the other hand, the principle seems well motivated. As in the case discussed above in which the order in which Alice and Bert arrived with their reports seemed irrelevant, it seems that it should not make a difference if I receive a piece of information in two stages, rather than one. Of course, it might make a *psychological* difference, but the issue is whether rational methodology puts different normative constraints on belief revision in the two cases. And, of course, it is always possible that the meta-information that I receive the information in two stages is relevant information. (My background knowledge might be such that the fact that some informant bothered to give me a partial report tells me something that is relevant.) But that does not seem to be what is going on in this example. If there is a rebuttal to this kind of example, I don't know what it is. Nor do I have any explanation for why difference between the one and two stage scenarios in this kind of case might be epistemically relevant.

Just to add to the puzzlement: there is a stronger version of (C1), implicitly rejected by DP, but proposed as one of a different set of revision postulates by Daniel Lehmann.[28] In the belief system notation, this is the principle:

(C1+) If $\alpha \subseteq \phi$, then $\Psi(\beta_1,\ldots\beta_n, \phi, \alpha, \beta_{n+1},\ldots\beta_m) = \Psi(\beta_1,\ldots\beta_n, \alpha, \beta_{n+1},\ldots\beta_m)$

The difference is that the stronger principle says that the whole belief system that results from the change, and not just the resulting belief state, should be the same in the two-stage procedure as it is in the corresponding one-stage procedure. The intuitive motivation for (C1) seems to apply to the stronger principle as well: why should it matter for your posterior belief revision policies whether you get partial information before getting the same whole information? But the stronger principle cannot consistently be added to the DP postulates.

The apparent counterexamples to (C1) are puzzling, but I think that there are clearer and more decisive problems for (C2). Consider this example: Fair coins are flipped in each of two rooms. Alice and Bert (who I initially take to be reliable) report to me, independently, about the results: Alice tells me that the coin in room A came up heads, while Bert tells me that the coin in room B came up heads, and so this is what I believe at stage one. Because my sources were independent, my belief revision policies, at stage one, will give priority to the H_AT_B and T_AH_B possibilities over the T_AT_B possibility (were I to learn that Bert was wrong, I would continue to believe that Alice was right, and vice versa).[29] But now Carla and Dora, also two independent witnesses, whose reliability in my view, trumps that of Alice and Bert, give me information that conflicts with what I heard from Alice and Bert. Carla tells me that the coin in room A came up tails, and Dora tells me the same about the coin in room B. These two reports are also given independently, though we may assume simultaneously.[30] This is stage two. Finally (stage three), Elmer, whose reliability trumps everyone else's, tells me that that the coin in room A in fact landed heads. (So, Alice was right after all). What should I now believe about the coin in room B? DP's postulate (C2) requires that I return to the original belief that the coin in room B came up heads.[31] Even though Dora's information had overturned Bert's information about the second coin, and even though Elmer provided no information at all about the result of the coin flip in room B (we may assume he knew nothing about it),

[28] Lehmann 1995.

[29] The full AGM system requires a total ordering of the possibilities, so the assumption that neither HT nor TH has priority over the other implies that they are tied. It might be more reasonable to assume that they are incomparable (and that the ordering of possibilities is only a partial order). But all we need for the current argument is that both HT and TH have priority over TT.

[30] I am assuming here that the conjunction of the two simultaneously received reports count as a single input, since the subject comes to believe, at one time, that both reports are true. One might question this assumption, but nothing about the theory as it stands provides any constraint on what counts as a single input, or any resources for representing the independence of sources.

[31] Here is the argument: Since H_A contradicts T_AT_B, $\Psi(H_AH_B, T_AT_B, H_A) = \Psi(H_AH_B, H_A)$ by (C2). But since H_AH_B entails H_A, $\Psi(H_AH_B, H_A) = \Psi(H_AH_B)$.

Elmer's information still forces us to change our belief about this result (if we follow the DP constraint, C2). This seems just as unreasonable as the result in the counter-example to Boutilier. In both cases, information is lost which, intuitively, should have been preserved.

One might object that I should respond as I did to Hans Rott's example, discussed above, by saying that relevant information from the intuitive story has been left out of the model. At each stage, I received the information, not only that the coins landed a certain way, but also that I received this information in some cases from independent sources, and that meta-information is clearly relevant to intuitive judgments about the way that beliefs should be revised. But in this case, adding this additional information would not help to avoid the consequences of the iteration principle. The full information that I receive from Elmer's report still contradicts the total information that I had previously received from the reports of Carla and Dora, so the argument still holds.

Unlike the situation with (C1), with (C2) it is not hard to see what goes wrong at a general level. (C2) directs us to take back the totality of any information that is overturned. Specifically, if we first receive information α, and then receive information that conflicts with α, we should return to the belief state we were previously in before learning α. But this directive is too strong. Even if the new information conflicts with the information just received, it need not necessarily cast doubt on *all* of that information. The previous information might have been quite complex, consisting of independent parts, and it might be that the new evidence threatens only one of those independent parts. To take an extreme case, just to dramatize the problem, suppose I start out knowing almost nothing, and then receive, at once, a fat encyclopedia of information. After I have digested it, I am told that there is a typo on p. 576: the Battle of Hastings was in 1066, rather than 1606. Should I then throw out the encyclopedia and go back to my prior state of almost total ignorance? That does not seem reasonable. The problem parallels the general problems of counter-factuals and of belief-contravening evidence. In all these cases the problem is to decide what to give up and what to keep when one must give up something to maintain consistency. This kind of problem arises, not only for my whole prior belief state, when the evidence conflicts with it, but also for the specific information just received, when the still newer information conflicts with it. As we have all known since Nelson Goodman gave up on the problem of counterfactuals more than sixty years ago, there is no general formula for answering this kind of question. This lesson gives us reason to reject (C2), despite the fact that we thereby lose the nice formal structure of the DP theory and of the Spohn iteration procedure, with its structural parallel with generalized conditionalization. If we do give it up, we will be left, at best, with a relatively weak set of constraints on iterated belief revision.

So where do we go from here? The examples we have discussed bring out the importance of various kinds of meta-information. In many examples, assumptions about the sources of our information, in particular about the independence of sources, played a central role in motivating intuitive judgments about how to repair our belief state in the face of evidence that something we took ourselves to know is in fact false. To decide how to revise in a particular case in which we learn that we were mistaken, we need to think about what went wrong, and how to explain it. In discovering that an item of presumed knowledge was in fact not knowledge, we discover that there was some defect or corruption at some point in the channels through which information was transmitted to us, or in the reasoning we used to reach a conclusion from the information we received. The more we know about the source of the problem, the better repair job we can do. We, as believers revising our beliefs, need to think about explanations, not only for the things that we take ourselves to know, but also for the fact that we know them, or in the case of putative knowledge that was overturned, the fact that we did not know them. And as theorists, we might be able to say something more substantive about this kind of process if we had the resources to represent meta-information, and its relation to ground-level information about the subject matter that is the primary focus. There are different kinds of independence—conceptual, causal, and epistemic—that interact, and one might be able to say more about constraints on rational belief revision if one had a model theory in which causal-counterfactual and epistemic information could both be represented. There are familiar problems, both technical and philosophical, that arise when one tries to make meta-information explicit, since it is self-locating (and auto-epistemic) information, and information about changing states of the world. But the problems are not intractable; a good theory of iterated belief revision will need to tackle them, and I would expect a theory that did to yield interesting results.[32]

[32] Thanks to the participants in the Kontanz workshop on conditionals and ranking functions, in particular to Thony Gillies, Franz Huber, Hans Rott, and Wolfgang Spohn, for discussion of the issues about iterated belief revision and ranking functions, and for advice about the relevant literature. Thanks also to the editors and an anonymous referee for helpful comments and corrections. I am particularly grateful to Hans Rott, who gave me detailed comments and suggestions on a previous draft, leading to some revision of my beliefs about belief revision, and I hope to significant improvements in the chapter.

4

Modeling a Perspective on the World

1. Introduction

John Perry's classic papers, published in the late 1970s,[1] convinced many philosophers that there is a special problem about self-locating belief (belief about who and where one is, and what time it is). Beliefs of this kind, it was argued, are not reducible to beliefs about the way the objective world is in itself. If *propositions* are items of information that are timelessly true or false, then one cannot explain a person's self-locating beliefs entirely in terms of the propositions the person believes. The most prominent arguments for this thesis involved cases of self-locating ignorance: cases where an agent knows the relevant objective facts, but remains ignorant of who or where he is. There was the amnesiac, Rudolf Lingens, reading a biography of himself in the Stanford library without realizing that the biography is about himself. He knows various facts about Lingens without knowing the corresponding facts about himself. Some examples are more mundane: I may know, at noon, that the meeting starts at noon without knowing that it is starting *now*. The most extreme of the examples of self-locating ignorance was David Lewis's story about two gods, one on the tallest mountain and the other on the coldest mountain.[2] They are both (according to Lewis) omniscient about the objective facts, but neither knows which the two gods he himself is.

I argued in an early paper on indexical belief that these ignorance cases were examples of a more general phenomenon that did not essentially involve beliefs about oneself or about the present time.[3] They were a special case of the familiar Frege puzzles in which a believer knows things about an individual in two different ways without realizing that these were ways of knowing about the same individual: the Morning Star and the Evening Star, Ortcutt, the man on the beach and Ortcutt, the man in a brown hat, the city called "Londres" in French, and the city called "London" in English, Paderewski the statesman and Paderewski the musician. I continue to think that cases of self-locating ignorance should be explained with the same resources used to explain the Frege cases, and in a parallel way, but I now

[1] Perry 1977 and 1979.

[2] Lewis 1979a. Lewis was not the first to worry about the indexical beliefs of omniscient gods. See Kretzmann 1966 for an argument that omniscience conflicts with immutability.

[3] Stalnaker 1981. Others at the time and since have made the same point. See Boër & Lycan 1986 and Cappelen & Dever 2013.

think that it is a mistake to conclude from this that there is no special problem of self-locating attitudes. Cases of ignorance about one's place in the world are just one manifestation of the distinctive role of one's perspective on the world, and one can show, without invoking such cases, that a purely impersonal and time-independent representation of an agent's state of belief (at a time) inevitably leaves something out.

Consider the following simple scenario: Albert is in the kitchen and Boris is in the basement. Each knows who and where he is, and who and where the other is, so there is no self-locating ignorance. They each know all the same objective facts about their respective locations in the house, but there is still a difference in their epistemic states, a difference in their perspectives on the world. To see that this difference is not reflected in their impersonal beliefs, consider any representation of the contents of their common state of belief: One might, as I would prefer, represent the contents of their beliefs with the set of the possible worlds that are compatible with the way they both take the world to be. But if you like, take the totality of their beliefs to be just a set of propositions, using your favorite account of propositions. All that matters for the point is that propositions are things with absolute truth conditions. That is, they are things whose truth conditions are invariant with respect to time, place, and person. Suppose such a representation contained all the information about the beliefs of any person who is in the cognitive state that Boris and Albert are both in. Let *x* be any person in that state. Where does *x* believe himself or herself to be? It is clear enough from the description of the scenario that Boris believes he is in the basement and Albert believes he is in the kitchen, but these are further facts that are not reflected in the common set of propositions that is what each of them believes, or in the set of possible worlds that are compatible with the way they take the world to be.

The point is not that Boris or Albert might be ignorant of which of the two he is. That may be true, but it is not part of the argument. The point is that our representation of their cognitive states is incomplete if we leave out the facts about the agent's perspective on the world, or what he takes that perspective to be.[4]

2. Centered Possible Worlds

David Lewis developed a very elegant framework for representing the additional information needed to specify fully a cognitive state—a modification of the possible-worlds representation.[5] A belief state, in his framework, should be represented by a set of *centered* possible worlds, rather than a set of simple possible worlds. A centered world is an ordered pair consisting of a possible world and a *center*, which is a spatio-temporal location in that world. (Alternatively, we could define a center as a pair consisting of a person and a time.) The center represents the time and place where the agent takes herself to be in the world as she takes it to be. I have followed Lewis in

[4] The simple point made by this argument will not be controversial among those who, like both Lewis and Perry, think there is a distinctive phenomenon of self-locating belief. The argument is directed at those, like Cappelen & Dever 2013 who argue that there is no problem of the essential indexical. The main point I will be arguing for is more controversial: that the distinctive character of self-locating attitudes does not imply that there is a distinctive kind of self-locating *content*.

[5] Lewis 1979a. See also Lewis 1983a.

using this kind of representation, but I have a different take on the job that is being done by the additional information about the agent's cognitive situation that the center provides. Lewis, focusing on the ignorance cases, took the problem to be that we need finer-grained objects to represent the contents of self-locating belief. The two gods case, for example, was supposed to be an instance of *intra-world ignorance* (to adopt a term used by Clas Weber[6]): ignorance that is witnessed by distinct centers within the same world.[7] Since the two gods are objectively omniscient, there is just a single possible world compatible with their knowledge. What they are ignorant of is where, within that single world, each is located. My explanation for why we need the richer representation of a cognitive state that Lewis's framework provides, and my account of the role that the centers are playing in that explanation, is different. What is needed to represent self-locating information is a way of linking the subject of a belief state (in a possible world in which she is in that state) to the places in the worlds compatible with that subject's beliefs that indicate that agent's point of view on the world. The centering is needed even in cases where there is no relevant self-locating ignorance (that was the point of the example of Boris and Albert). Furthermore, I have argued that even in cases of self-locating ignorance, there is no such thing as intra-world ignorance.[8] Ignorance about where one is in the world is *always* also ignorance about which possible world is actual. If one is mistaken about where or who one is, or about what time it is, one is mistaken about what the world is like. I will follow Clas Weber in labeling this thesis (which he argues against) "*Propositionality*." On this hypothesis, the case of the two gods cannot really be a case of omniscience about what the world is like in itself. The ignorance is witnessed by a pair of possible worlds in which the gods change places. If one of the gods does not know whether he is on the tallest or the coldest mountain, then there is an epistemically possible world where he is on the one, and a different epistemically possible world in which he is on the other. All cases of apparent intra-world ignorance or agnosticism can be treated in a similar way, or at least that is what I have argued.

Lewis represents the *contents* of belief, self-locating and otherwise, as sets of centered worlds, or equivalently as *properties*, which are abstracted away from the particular agent who has the belief. So, on his account, the contents of my beliefs about myself have nothing to do with myself. (Of course, if I believe that my pants are on fire, I am ascribing a property—having pants that are on fire—to myself. But the content of my belief is just the property, and *that* is not something that, in itself, involves me.) If we accept propositionality, we can represent the contents of belief, self-locating and otherwise, as ordinary timeless and impersonal propositions: the contents of my beliefs about myself will be themselves about myself. This way of representing content facilitates the comparison of the beliefs of different people, and

[6] Weber 2015.

[7] In the representation of a knowledge or belief state by a set of worlds or centered worlds, for any proposition P, there will be a (centered) possible world in which P is false in the set if and only if the agent in that state does not know or believe that P. Ignorance of whether P is "witnessed" (as I am using that word) by a pair of (centered) worlds, if and only if P is false in one and true in the other, and they are both in the set that is being used to represent the knowledge or beliefs of an agent who is in the relevant state.

[8] In Stalnaker 2008, 2011b, and 2014.

the representation of the way a person's beliefs change over time. This was the main motivation I offered in defense of *Propositionality*, but there are other ways of modeling belief dynamics and the communication of self-locating information, and the hypothesis remains controversial. My aim in this chapter is to develop my defense of the hypothesis in response to some criticisms, emphasizing broader issues concerning the explanation of intentionality that I think are at stake. My view is that *Propositionality* is a plausible thesis, given a certain externalist picture of the way the facts about a person's situation in the world determine the content of his or her cognitive states, and that resistance to that hypothesis rests on a contrasting internalist picture, such as the one that figures in David Lewis's metaphysical theory. It is the opposing general pictures that are of interest. The hypothesis about intra-world ignorance is just a way of bringing out those issues. I will focus on some arguments in a recent paper by Clas Weber, since his discussion connects the thesis of Propositionality with the broader issue, particularly with the role of singular propositions in the representation of the contents of belief.

3. Against Intra-World Ignorance

One has a case of intra-world ignorance if one has a state of belief that is modeled, in the centered-worlds framework, by a set of centered worlds that contains at least two that differ only in the location of the center. There will be such cases if and only if there are situations in which there are possible worlds compatible with the beliefs that a certain agent has at a particular time that contain one or more other agents with exactly the same *self-conception* at a corresponding time, where a person's self-conception is the set of all beliefs that the person has at the time about him or herself. Weber acknowledges that for ordinary believers in ordinary situations in which they are not sure where they are, it might seem unlikely that there are such possibilities:

> My self-conception is rather rich and includes beliefs like *I'm called "Clas," I'm currently eating a sandwich, having such-and-such sense-experiences and memories, etc.* To suffer from individual intra-world ignorance [about, for example, whether I am in Canberra or Perth], there would have to be a doxastically accessible world in which there are at least two persons with my name, both eating a sandwich, having exactly the same sense-experiences and memories, etc. one located in Canberra, the other in Perth.[9]

But he argues that even if the existence of more than one agent with the same subjective perspective on the world would be a bizarre and unlikely hypothesis in most situations, one cannot exclude even realistic examples of this phenomenon. First, some agents will have a much more impoverished self-conception than normal people have in normal situations, and these self-conceptions might be more plausibly duplicated. Second, even if it is highly unlikely that such scenarios are actually realized, it is not so unlikely that there are real agents (perhaps skeptics or proponents of a hypothesis of eternal recurrence) whose beliefs are at least compatible with

[9] Weber 2015, 664.

such a scenario. While I am not convinced that there are realistic examples of worlds where agents have subjective duplicates, I agree that one cannot exclude such scenarios on grounds of implausibility, and I never meant to suggest otherwise. Brains in vats, evil demons, swampmen, and twin-earth stories are contrived and implausible too, but that observation is not a response to any of the arguments in which those thought experiments play a role.

To nail down the possibility of subjective duplicates, Weber considers the hypothesis that we live in a symmetrical universe, with a pair of agents who are indiscernible, not only in their subjective perspective, but in all qualitative ways. I don't have any objection to the assumption that such a hypothesis is coherent, and that it may be epistemically possible for some real people. My central argument for the hypothesis of Propositionality was that we don't need to exclude the duplication of subjective perspectives in order to exclude intra-world ignorance. The claim is that even on the hypothesis that we live in a symmetrical universe, with a distinct region indiscernible from this one, an agent (whose beliefs allow for this possibility) can distinguish herself from her twin, and her region from her twin's region.

Suppose Agnes is discussing her hypothesis that we may live in a symmetrical universe with her colleague Bert. I assume that it is part of Agnes's self-conception at that time that she is talking to Bert. According to the hypothesis that she is floating, there is or may be in a remote part of the universe a twin-Agnes talking, at the counterpart time, to a twin-Bert. It is part of twin-Agnes's self-conception that she is talking with twin-Bert. But he is a different person from Bert, so twin-Agnes's self-conception is different from Agnes's. This is, therefore, not a case of intra-world ignorance.[10] Or perhaps the hypothesis Agnes is floating is that a planet something like Putnam's Twin Earth exists in a remote part of the universe. Agnes is drinking a glass of water, and is aware of the fact, as she discusses the hypothesis, so it is part of her self-conception that she is then drinking a glass of water, while Twin-Agnes's self-conception has her drinking twater (though she calls it "water"). Agnes need not know any chemistry to distinguish Earth from Twin Earth. All she need assume is that twater is a different kind of stuff from water. Assuming Agnes is sufficiently ignorant of chemistry, perhaps the twins are phenomenally alike, but they still have different self-conceptions, on what I think is the right way to understand intentional states.

Suppose Agnes has a particularly impoverished self-conception: she is an amnesiac alone in a fog, not even aware of the fog, but still able to entertain the symmetrical-world hypothesis. Even in this situation she can distinguish herself from her twin on the other side of the universe. She herself is *here*, as she might put it to herself, while her twin is not. She recognizes that her twin will be thinking a parallel thought, using the same demonstrative with the same meaning, but the reference of the demonstrative will be different, and recognized by both to be different. One might question

[10] Weber considers the possibility that one might use other things in one's environment to distinguish one's self-conception from that of one's twin, but argues that even if this were to work, one can't account in this way for some cases of ignorance of the present time. More generally, he argues that this strategy involves a problematic commitment to singular propositions. I will consider time reference and the more general issues below.

whether this difference in reference is enough to make a difference in self-conception, but if the difference between Bert and twin-Bert, in the first case, is enough to distinguish Agnes's self-conception from that of her twin, why not the difference between Agnes herself and her twin?

The claim that we can distinguish the self-conceptions of Agnes and her twin is not based on the assumption that she has some nontrivial cognitive capacity to distinguish herself, in qualitative terms, from everyone else in the world. The claim is that a self-conception has a demonstrative component. The reason this is important is that the agent's particular location in the world provides the anchor that determines the way all of her thoughts get their content. Agnes can think about Bert, Aristotle, Uzbekistan, the Taj Mahal, the planet Neptune, Balzac's novels, water, giraffes, and dinosaurs, the colors pink and green, genes, gross national products, Baroque churches, and herself, only by being in an environment, broadly construed, that contains these things. The space of possibilities that represents her knowledge and ignorance about that environment can be represented, by her or by others, only in terms of the things, properties, and relations with which she interacts. The times and places where things are located in the possible situations that define that space must be calibrated, one way or another, with the time and place she finds herself in in the actual world, and with the other times and places she believes to characterize that world.

If one gives nonzero credence to a symmetrical universe hypothesis, then one allows that one might have a subjective duplicate somewhere else in the universe, but that does not by itself imply that that one is ignorant, in any sense, about where one is in the world. Even if Agnes assumes the truth of this hypothesis, she cannot be ignorant of which part of the universe she is in unless she has some neutral way of identifying the alternatives. If all Agnes can say is that her subjective duplicates are somewhere else, then there is no ignorance, since she knows that she is not one of the ones who are somewhere else. But suppose she does have an independent way of identifying the different cases. Suppose, for example, that according to the symmetrical-world hypothesis that Agnes is entertaining and giving nonzero credence, the parallel parts of the universe are not simultaneous.[11] In this almost-symmetrical world, one sequence of events is taking place ten minutes earlier than the corresponding duplicate sequence, but Agnes has no opinion about whether she is in the earlier or the later sequence of events. Now we have a genuine case of ignorance about where in the world Agnes is, but not thereby a case of intra-world ignorance. It would be natural to describe the case in terms of two alternative doxastically possible universes, only one of which is actual. In one of them, there is a duplicate sequence of events taking place ten minutes earlier than this one, and in the other, there is a duplicate sequence that takes place ten minutes later.

In this story, is Agnes ignorant about what time it is? Not obviously, assuming that her watch is working, and that she consults it. She notes that it is 2 a.m. as she hears a loud thunderclap, and reflects that (on the condition that her symmetrical-universe hypothesis is true) her twin either heard a corresponding thunderclap at 1:50, or will

[11] I hope Agnes and I can be forgiven for assuming that she lives in a Newtonian world.

hear one at 2:10. So she knows what time it is, but not what time it was or will be when her twin has the parallel experience. But if Agnes is perverse, she might use the time zone of her twin to measure the relevant times. Since Agnes knows that her twin's watch, if accurate, will read 2 a.m. when she (the twin) hears the thunderclap, Agnes knows that when her (Agnes's) watch says 2 a.m., it is "really" (parallel-world Standard Time) either 1:50 or 2:10, but she doesn't know which. Which of the two ways of fixing the reference of the time itself is the right one? One seems more natural for Agnes, the other for her twin, but there does not seem to be a factual question here, even on the assumption that the symmetrical-world hypothesis is true.

Symmetrical universes are not needed for subjective duplicates, and Weber does not suggest that they are. They are introduced just to nail down the possibility. More generally, the environments of subjective duplicates can be very different, so long as the differences do not impinge on the subjective experiences of the corresponding agents. Weber's admittedly implausible but still possible case of two people named "Clas" with the same sense experiences and memories, one in Perth and the other in Canberra, was far from a symmetrical-universe hypothesis, as is Lewis's notorious case of the two gods. The objective differences between the subjective duplicates make it all the easier to give a neutral characterization of the different epistemic possibilities. Clas (according to the story) does not know whether he is in Perth or Canberra, but he can distinguish Perth from Canberra in a way that is independent of the fact that one of them is where he is and the other is where he isn't but might (for all he knows) be. The propositionalist's strategy, in all these cases, is to represent the agent's ignorance about his self-location with distinct possible worlds. If Clas is really in Canberra, with a twin in Perth, and does not know which of the two places he is, there is (according to the propositionalist) a non-actual possible world in which he himself is in Perth. The epistemic possibility of a twin within the *same* possible world is in any case irrelevant (at least in one natural elaboration of the story) to his ignorance of his own location, as is shown by the following twist on the story: suppose a trusted and trustworthy oracle informs Clas that he has no subjective duplicate. "At no place or time in the actual history of the universe," the oracle reports, "is there anyone who is indistinguishable, from the inside, from you." Clas learns something about the world from this report, but he does not learn anything about his own location. In particular, he is just as much in the dark as before about whether he is in Canberra or Perth. So, to represent Clas's beliefs after getting the oracle's message, we will need a nonactual possible world in which he is in Perth rather than Canberra. But that epistemically possible world must have been there all along.[12]

[12] If we assume that Clas did not positively believe that he had an actual subjective twin, but only allowed this as a doxastic possibility, then the oracle's message added to his beliefs (and so eliminated some prior doxastic possibilities), but did not contradict any prior beliefs (and so did not add any new doxastic possibilities). But I do not mean to claim that a belief in subjective duplicates, or in the possibility of subjective duplicates, is *never* relevant to doubts about one's location in the world. Suppose Clas knew that *someone* exactly like him, subjectively, was in Perth, but didn't know whether that person was himself, or someone else. Then the oracle's message would tell him that he must be in Perth. More generally, if one's reason for being in doubt about where one is depends on a belief that one might have subjective duplicates,

Or consider the two gods. It is hard for an oracle to give new information to omniscient beings, but suppose just one of the gods (say the one on the tallest mountain) learns which of the two he is—that he is the god on the tallest mountain. "I've been informed," he announces to the other god, "which mountain I am on, but I was sworn to secrecy by the oracle who told me, so I am afraid I can't help you learn who and where *you* are." We no longer have a case that is even a candidate for intra-world ignorance, but the unfortunate god on the coldest mountain remains as clueless as he was before about whether he is on the tallest or the coldest mountain, and whether he throws down manna or thunderbolts.[13] We now *must* represent the ignorance of the god that remains ignorant with a pair of distinct possible worlds, one in which the god on the tallest mountain is the clueless one, and one in which it is the god on the coldest mountain. But if we had initially represented the ignorance with a pair of distinct possible worlds in which the gods traded places, then we can represent this change in the epistemic situations of the gods in a straightforward way: In the prior state, there were two possible situations compatible with the knowledge of both gods. Then one of the gods acquired information that ruled out one of those two possibilities, while the other did not. The gods also have, both before and after the oracle's intervention, some knowledge about what they each know and don't know: In the prior state, it was common knowledge that neither god knew which he was, while in the posterior state it is common knowledge that one of them knows, and the other does not.

One should not rest very much on the case of the two gods, or on modified versions of it, since these stories are of doubtful coherence, but the dialectic of argument about the right way to understand the epistemic situations of the gods is instructive, so let me consider David Lewis's reasons for rejecting the representation of their ignorance by two possible worlds with the two gods switching places. Lewis has two reasons for not liking this representation. The first reason is metaphysical: Lewis is an anti-haecceitist who believes that if two possible worlds are qualitatively indiscernible then they are indiscernible *tout court* so that it makes no sense, given his metaphysical assumptions, to suppose that there are distinct but qualitatively indistinguishable worlds in which particular individuals are interchanged. Individuals, on Lewis's metaphysics, exist only in one possible world; they have counterparts in other worlds, but the counterpart relation must be explained in terms of qualitative similarities and differences between the individuals. Lewis's anti-haecceitism is motivated by his modal realism, but is separable from that metaphysical doctrine, and might be shared by an actualist. It is also motivated by Lewis's internalism about intentionality. I agree that my favored way of modeling this case and others

then the doubt will be resolved if one learns that one does not. (Thanks to Stephan Torre for pointing out the need for this clarification.)

[13] I am playing fast and loose with Lewis's story here, and this variation on the story is not an argument against his account of his original case. The omniscience that Lewis assumed must include omniscience about the future course of events in the world, and that is not compatible with my variation. If the gods had known, before the event, that an oracle would later give private information to one of them, but not the other, then they would not have been omniscient, at the earlier time, about even the qualitative features of the world.

presupposes a notion of *primitive thisness* (to borrow a phrase from Robert Adams) but that is a commitment I am happy to make.[14]

Lewis's second reason is more interesting, since it brings out the connection with the general issues about intentionality. He argues that the haecceistist's response won't solve the problem anyway, even given this metaphysics, for the following reason: Suppose, he says, there really are two qualitatively indistinguishable worlds, with the gods interchanged. Call the two worlds W and V. Suppose an oracle inform both of the gods that they are in world W. They will then be truly omniscient about the objective facts, but Lewis argues that this would still leave them ignorant about their location in the world. But let's look more closely at how the gods are informed about which of the two possible worlds they are in. A state of knowledge or belief is not given just by specifying the possible worlds compatible with it. One needs to say how the facts about the agents and their situation in the world make it the case that the given set of possible worlds is a correct representation of their knowledge or beliefs.

Suppose our gods have names: one is Castor and the other is Pollux. W is the world in which Castor is on the tallest mountain and Pollux on the coldest mountain, while V is the world in which their positions are reversed. But how did the gods get their names, and what do the gods know about who those names refer to? One way to fill out the story is this: The sophistical oracle says, "Let 'Castor' be a rigid proper name for the god on the tallest mountain in one of the worlds (call it 'W') and let 'Pollux' be a rigid name for the other god. Call the world where their positions are reversed 'V'. The actual world is W." Granted, this is no help in removing the self-locating ignorance, but the gods rightly complain that they have been given no information at all, but just a tight circle of stipulations. "We didn't need you for this exercise," they protest. "We could have made those stipulations ourselves, taking 'W' to be a name for the actual world. You didn't tell us what world is actual any more than Gareth Evans told his readers who invented the zip when he stipulated that 'Julius' shall be a rigid designator for the person who did, and then told them that it was Julius who invented the zip."[15] As the analogy with Julius brings out, the fact that this kind of exercise conveys no real information has nothing special to do with self-location.

A better story is this: "Let's give ourselves names," says one of the gods. "I'll be Castor, and you can be Pollux." Now we have an impersonal way of stating what neither of us knows, and we can ask the oracle to give us this objective, impersonal information. Is the actual world W, where Castor is on the tallest mountain, or is it V? If the oracle answers *this* question, there will be no more self-locating ignorance. The gods' reference-fixing stipulations do not, of course, give themselves any information, but they do succeed in transforming a question into one that can be stated as a question about what the actual world is like, in itself. The upshot is, as one should expect, given the thesis of Propositionality, that questions about where one is in the world just are questions about what the world is like in itself, asked (as all questions are) from a particular perspective.

"Not so fast," the Lewisian replies. "Remember that while the gods are agents who do different things (one throws down manna while the other throws down

[14] See Adams 1979 and Stalnaker 2012.

[15] Evans 1979.

thunderbolts), our story requires that each remains ignorant, as he is acting, about whether it is he himself or the other one who is the agent of the action." (This is one of the sources of conceptual strain in the story, but suppose we can make sense of it.) "This point applies to speech acts, including reference-fixing stipulations. What this implies is that if it is the god on the tallest mountain who said, 'I will be Castor,' all Castor learns is that the god on the tallest mountain stipulated that his name should be 'Castor'. He does not know whether it is he himself who has that name since he does not know whether it was he himself who made the stipulation. So, this variation on the story of the oracle's way of giving the gods information about which of the worlds is actual is no better than the first." Point taken, but the upshot is not that the oracle can tell the gods which of the two worlds they are in without removing their self-locating ignorance; it is rather that (given the problematic supposition that agents may be alienated from themselves, their actions, and their world in this radical way), the oracle is incapable of telling the gods which of the two indiscernible worlds they are in, at least in a way they will understand. What the dialectic brings out is that this upshot depends on the distinctive problematic feature of the two gods' case and will not carry over to any situation in which it is assumed that an agent knows, when he speaks, that it is he himself who is speaking. The maneuver of naming oneself, and so turning information specified with a personal pronoun into information that can be specified in impersonal terms will be available in any case that meets this modest condition.

So, let us leave the two gods behind. Problems remain, even if we restrict ourselves to situations that lack the peculiar and dizzying kind of ignorance suffered by the epistemically challenged, if omniscient, Lewisian deities. Weber argues that my way of avoiding intra-world ignorance, because of its appeal to singular reference and its reliance on singular propositions, faces a dilemma, and that either way of responding to it will fail to give an account that answers to my own motivations. His discussion brings into focus some tensions in the use of singular propositions to describe the doxastic and cognitive states of agents, but I am not persuaded by his specific arguments, since they depend on premises that I don't accept. In the concluding section I will respond to his dilemma.[16]

4. Self-Location and Singular Propositions

My defense of *Propositionality* requires the assumption that reference to particular individuals plays a role in the identification of states of belief, and in the distinctions we make between them. I said that a person's self-conception had a demonstrative

[16] In the course of his discussion, Weber raises some more general problems about the use of singular propositions, in the context of a possible-worlds framework, to characterize an agent's cognitive situation. He claims, for example, following David Lewis, that beliefs with singular propositions as their content are "extremely strong": "Having a singular belief about an object entails having true beliefs about its essence" (p. 13), and argues that this is not compatible with the assumption that ordinary beliefs about people and places in our environment have singular propositions as their contents. This problem needs to be addressed, as do the familiar Frege puzzles in which a person or city is known in different guises, but they are not specific to the problem of self-location. See the papers reprinted in Stalnaker 1999a, and especially Stalnaker 1988, for my strategy for responding to these problems.

component, and this is essential to the defense, but I do not assume that beliefs expressed with demonstratives are always singular propositions. If I say, "I wonder who that is?" when I hear the doorbell, I refer demonstratively to the person who rang the bell, but I don't thereby have singular beliefs about that person, if by a "singular belief" we mean a belief whose content is a singular proposition. In most normal cases, the first-personal beliefs I express with the word "I" will be singular propositions about myself, but not always, and usually not in the puzzle cases that are the center of attention in the discussions of the problems of self-location. Singular reference and singular propositions are part of the story, and we need to address some puzzles about them, but before getting to that let me respond to some particular arguments that I think do not raise real problems.

Weber attributes to me a thesis that he calls *Singular Self-location*: "Self-locating beliefs are nothing but binary relations to singular propositions."

I am not sure what this means, but I can think of no interpretation of it that is a consequence of the view I am defending. Let me sketch the picture of the way I want to model states of belief, and then try to interpret the thesis.

A belief state, we all can agree, is the belief state of a particular individual at a particular time in a particular possible world. If we represent belief states by sets of centered possible worlds, we can distinguish the possible world (centered on a particular person and time) *in* which that person is in a given belief state at the time from the possible worlds (also each centered on a particular person and time) that represent the belief state by being the set of ways the world might be, according to that person at that time. The traditional way of modeling states of knowledge and belief, going back to Jaakko Hintikka's early work on epistemic and doxastic logic,[17] is with a binary relation between possible worlds: The relation holds between world *x* (call it the base world) and world *y* if and only if world *y* is compatible with what the believer believes in the base world, *x*. On the revised picture the doxastic accessibility relation holds between centered worlds. The center in the first term of the relation—the base centered-world—represents the person whose beliefs are being modeled, and the time at which the person has those beliefs. The centers of the centered-worlds in the second term of the relation represent the person that person takes herself to be in a world that is compatible with the way she takes the world to be, and the time in that world that she takes it to be. The only difference between my version of this picture and Lewis's is that I impose the condition *Propositionality* on the set of doxastic possibilities: no centered worlds in this set with the same world, but different centers.

Now given this picture, what exactly does Weber's thesis, *Singular Self-location*, say? Here is one way to interpret it: first, for any belief state, represented this way, we can define the set of singular propositions about the person *a* who is in the belief state—the person at the center of the base centered-world. Then the thesis might say: *a*'s self-locating beliefs (at the time, in the base world) are the singular propositions about *a* that *a* believes at that time. This thesis is clear enough, but it is one that I definitely reject. It is compatible with the model for representing knowledge and belief that I have sketched, and central to the phenomena it aims to clarify, that a

[17] Hintikka 1962.

person *a* may be ignorant or mistaken about who he himself is, and about what time it is. A model that represents *a*'s beliefs in a case of this kind will be a model in which the persons and times at the centers of the worlds compatible with *a*'s beliefs at time *t* will in some cases be different from *a* and *t*. It is thus compatible with the picture that the agent *a* believes that he himself satisfies a certain condition without believing the singular proposition that *a* satisfies the condition, and also possible that *a* believes the singular proposition that *a* satisfies the condition while not believing that he himself does.

Here are the two horns of the dilemma that Weber poses:

> On the one hand, one may assume that access to singular propositions is easy. This assumption turns out to be inconsistent with *Singular Self-location*. Alternatively, one may hold that access to singular propositions is hard. This will force us to abandon *Transmissibility* [the thesis that a person's self-locating information can be communicated to others].[18]

The assumption of the second horn of the dilemma is the Russellian view that one can grasp singular propositions only about oneself, or about one's current sense data. I agree that this hypothesis is incompatible with the communicability of singular propositions. The assumption of the first horn is a thesis Weber calls *Easy access*: "There are ordinary, third-person ways of acquiring singular propositions about other subjects (e.g., through perception, testimony, etc.)."[19] I accept the thesis of Easy access, but the picture I am defending rejects the assumption that sets the dilemma up: *Singular Self-location*.

Weber gives two arguments for the claim that *Easy access* conflicts with *Singular Self-location*. The first is that since *Easy access* implies that one can have singular beliefs about others based on perception and testimony, one should also be able to have singular beliefs about oneself from a third-person perspective (as when one sees oneself in a mirror without realizing that it is oneself that one is seeing). I agree this is a problem for *Singular Self-location*, as I have interpreted it, but not thereby a problem for the picture I am defending.

The second argument is for the incompatibility of *Singular Self-location* with the assumption that two people can believe the same singular proposition.[20] This argument is a non-starter, even given *Singular Self-location* as I have interpreted it. The problem is that the argument takes it to be a consequence of that thesis that if a given proposition is self-locating for one believer, then that same proposition must be self-locating for anyone who believes it. But being self-locating (even assuming *Singular Self-location*) is not a feature of a proposition believed. Even in cases, such as the case of Boris and Albert, where there is no self-locating ignorance, and so Boris's self-locating beliefs are exactly the singular propositions about himself that he believes, those same propositions are also believed by Albert, but they are not self-locating for him.[21]

[18] Weber 2015, 653. [19] Ibid. [20] Ibid., 655.

[21] Weber has an argument against the proposal to treat self-location as a feature of the believer's relation to the content of the belief, rather than as a feature of the content itself. It begins with the following thesis, with which I agree, at least on one way of interpreting it: "It is the defining theoretical role of content to reflect differences in cognitive significance. Beliefs that represent things differently should be assigned

So, I don't think these arguments work. While I accept the *Easy access* thesis, it does not play a direct role in the defense of *Propositionality*. The defense of that thesis does depend on the possibility of singular reference, and reference-fixing, and the general framework presupposes the intelligibility of the notion of a singular proposition, assuming that what we mean by a singular proposition is just a proposition that is true if and only if a certain particular individual satisfies some condition. But we can defend *Propositionality* without making any assumptions about easy or hard access to singular propositions. What we need to assume is just that we have the capacity to characterize—to uniquely describe—certain singular propositions, as Gareth Evans did when he stipulated that "Julius" shall be a rigid proper name of the inventor of the zip.[22] The propositions that Julius invented the zip, that he was born in Minsk, that he also invented the light bulb, etc. are all well-defined singular propositions (assuming that there is a unique person who invented the zip). It is a separate question whether we are in a position to believe, or even to entertain them. If we can do what Evans did with Julius, we can also define singular propositions by using demonstratives or personal pronouns, again without making any assumptions about whether the person giving the definition has access to the proposition he is defining.[23] So, consider poor Clas, in Canberra, but not knowing whether he is in Canberra or Perth. As you recall, he gives nonzero credence to the proposition that he has a subjective duplicate in the other place. But now he says, "Call me 'Clas': as I am now using this name, it refers uniquely to me. Let me dub my subjective duplicate, should he exist, 'Twin-Clas' (though I recognize that he, confusingly, calls himself by the same name as I call myself). Now I don't know whether 'Clas is in Canberra' expresses a truth, nor whether 'Clas is in Perth' does. I do, however, know that there is no possible world in which the propositions they express are both true."

In this particular example, I think it is reasonable to say that the singular propositions that Clas is in Canberra and that he is in Perth are propositions he can grasp, suspend judgment on, assign credence to. But one need not assume that for the reference-fixing exercise to work. Most of the time most of us know who we are, but not always, and not necessarily. And in the case of reference to time, the corresponding sociological generalization is not even close to true. That is, most often

different contents" (Ibid., 658). It is then argued that since it makes a difference to the cognitive significance of a thought that it is self-locating, this feature must be built into the content. But I take the thesis that is the premise of this argument to concern the comparison of the cognitive significance of two beliefs of the same person at the same time. It implies that if O'Leary believes that O'Leary was born in California, but not that he himself was, then we must distinguish the content of the belief from the content of what he does not believe—that he himself was born in California. That is, the thesis implies that we should not explain the difference as a case where the same proposition is believed in one way, but not in another. But the thesis does not imply that the content of Boris's belief that he himself is in the basement must be distinguished from the content of Albert's belief that Boris is in the basement. These are not beliefs that have a different cognitive significance for some one believer.

22 Evans 1979.

23 I am here borrowing a device introduced by Sarah Moss in a discussion of updating of belief states. In one of Moss's examples, the subject, who may or may not know who he himself is, dubs himself "Dr. Demonstrative," and so knows a priori that de se propositions expressible with sentences of the form "I am F" are materially equivalent to de dicto propositions expressed with sentences of the corresponding form "Dr. Demonstrative is F". See Moss 2012.

most of us *don't* know what time it is, at least with any precision. Nevertheless, we can easily fix the reference of a name for the present moment. "Let 'T_0' name *this* time—the time it is *now*. I know that 'the meeting starts at T_0' is true if and only if the meeting is starting now." This exercise, like Evans's, is of course not a way of coming to know any substantive piece of information. If "T_0" in fact picks out noon, and I know that the meeting starts at noon, then I *do* know the singular proposition that the meeting starts at T_0. What I don't know, and can't learn simply by making a stipulation, is that "the meeting starts at T_0" expresses that proposition.[24] The sentence expresses different propositions in different possible worlds that are, for me, epistemic possibilities. But if my stipulation succeeds in determining a timeless proposition that is true if and only if a certain self-locating proposition is true, then whether or not I know what proposition that is, it will make the point that any pair of doxastic alternatives in which "now" picks out different times will be alternatives that differ with respect to the truth of a timeless proposition, and that is enough to establish *Propositionality*.

The reference-fixing exercise gives us an artificial Frege case: a case where one has access to the same thing (in this case a time) via two different causal routes. The demonstrative way of identifying a time is not that different from the demonstrative way in which one might have fixed reference to a planet. "Let *that* be Hesperus," I say, pointing to a certain planet in the evening. The next morning I point to a planet, or to something I suspect is a planet, and say, "I wonder if *that* is Hesperus too." Or perhaps I dub it "Phosphorus" and wonder whether Hesperus is Phosphorus. Of course, more realistically, I don't fix the reference of "Hesperus" and "Phosphorus" by stipulation—I use established names whose reference was determined by a robust and stable historical practice to which I defer. Here is a better analogy: Not being an experienced star-gazer, but knowing as every philosopher does that Hesperus is a planet that appears in the evening (and the morning) I scan the night sky and wonder as I point; "Is *that* Hesperus, or is it *that*?" There is no need to give temporary names, on the fly, to the things I point to, but I could. My "*that*"s, or the names I use them to fix, are anchored to the situation in which I use them, but they pick out different celestial objects in the different epistemically possible worlds that represent my cognitive situation at the time.

The reference-fixing maneuver is just an artificial device for making a philosophical point, but is also a way of bringing out the possibility of calibrating the time references at different times. Even if I don't know what day a particular token of "today" refers to, I or someone else may be able to refer, anaphorically, to that day at a different time. If the next day I remember the earlier time, I can say that I didn't know then what day it was then. I might also say that I still don't know (that same piece of information). Even if I still don't know what day that was, I might know that

[24] The *diagonal* proposition determined by that sentence, relative to the epistemic possibilities compatible with my beliefs, is the proposition that is true in the possible world if and only if the propositions expressed in that world are true. My general strategy for treating the Frege cases, and other problems about singular propositions of the kind that Weber raises, has been to take the diagonal propositions, in problematic cases, to be the content of belief. (A diagonal proposition is not a particular kind of proposition, but a particular way of determining one.) See the papers in Stalnaker 1999a.

the earlier time (which was in fact yesterday) was yesterday, or alternatively, I might recall my earlier situation without remembering whether it was yesterday or the day before. In this case I am in doubt, both about the objective reference of my past "now" and current "then," and also about the relation between that time and *this* one (my current "now"). But I do know that that past "now" referred to the same time as my current "then." Attributing, at one time, attitudes that I or another person had at another time is a matter of describing a person's perspective on the world from a different perspective. To do this, we use the resources available from our current perspective to characterize both the location of the other perspective, and the information that defines the other cognitive state. We connect our "now"s and "then"s to each other, as well as to the objective time frame in order to describe and explain our knowledge and ignorance at one time about our knowledge and ignorance at other times and to represent learning and forgetting and change of mind.

In a conversation, two people can connect each other's "I"s and "you"s, even if neither knows (in a sense) who he is talking to. Lingens the amnesiac may be talking to Daniels (though he didn't catch his name) about who he (Lingens) might be. Neither Daniels nor Lingens knows of Lingens that he is Lingens, but they do know that the person one of them refers to with "I" is the same as the person the other refers to with "you," and that is enough to create a context that makes communication possible.[25] There may be ways of explaining the way we calibrate cognitive states of different people and at different times that don't depend on the thesis of Propositionality, but this assumption does help to give a smooth representation of such situations, and I think it is also well motivated in itself.[26]

[25] In Stalnaker 2011a, a revised version of which is chapter 5 of Stalnaker, 2014, I spell out a formal representation of *common ground* that provides a context in which self-locating communication can take place.

[26] I am grateful to Clas Weber for his paper, Weber 2015, which helps to clarify and sharpen the issues, and for comments on this chapter, as well as discussion with him about self-location. Thanks also to the editors and a referee for comments and corrections.

5

Reflection, Endorsement, Calibration

1. Introduction

Bas van Fraassen's reflection principle has come in for a lot of criticism since he first proposed it in 1984.[1] Counterexamples abound, and the counterexamples are both robust and diverse. Everyone agrees, including van Fraassen himself, that the principle, if defensible at all, requires serious qualification, but with suitable qualifications, as well as some generalization of the principle, I think it can be defended, and that it can do some work in explaining both rational belief change, and rational debate between different agents. I am going to start by looking briefly at the stock range of counterexamples, and then try to motivate the kind of qualification that I think is needed. But my main concern is not to defend the principle. I am interested in exploring a more general issue that the reflection principle helps to bring out: the role in inductive reasoning, belief change, and rational discussion, of an agent's doxastic attitudes (at a particular time) about his or her doxastic state at a different time, and about the doxastic states of other agents. It is obvious that any attempt to assess or justify one's beliefs, to consider how to respond to new evidence, or to discuss with others what it is reasonable to believe will involve consideration, from one's present epistemic perspective, of different epistemic perspectives, both one's own points of view at different times, and the points of view of others. Such reasoning and discussion will inevitably involve the attribution at one time of the knowledge and credences that characterize a different epistemic perspective. Knowledge and credence are characterized in terms of sets of alternative possibilities (epistemic alternatives), and measures on such sets. Attributing knowledge and credence to others requires characterizing, from one perspective, the epistemic alternatives that are available from a different perspective. Comparing a different epistemic situation with one's own requires *calibrating* the possibilities, as seen from one perspective, with the possibilities as seen from a different perspective, and this process of calibration is not always straightforward.

Here is the plan: first, I will state the reflection principle in its original unqualified form. Second, I will review some of the standard counterexamples, and propose two qualifications that I think are sufficient to make the principle defensible. But third, once the principle is qualified it can be generalized, and I will state a more general version of the reflection principle. But even with the qualifications I will suggest, the

[1] van Fraassen 1984. See also van Fraassen 1995 for a later defense, with responses to some criticisms.

principle will work only if we make a transparency or *luminosity* assumption. So fourth, I will briefly sketch the reason that this assumption is required, and defend it (or at least I will assert that it can be defended). Finally, I will look at the role of the generalized reflection principle, and of the general issue of the ascription at one time of credences at a different time in the dialectic of argument about a puzzling example.

2. The Unqualified Principle and Some Counterexamples

In "Belief and the will" van Fraassen proposed what he called "a new requirement of rationality, in addition to the usual laws of probability calculation":

(Reflection) $P^a_t(A \mid p^a_{t+x}(A) = r) = r$

"Here P^a_t is the agent a's credence function at time t, x is any non-negative number, and $(p^a_{t+x}(A) = r)$ is the proposition that at time $t + x$, the agent a will bestow degree r of credence on the proposition A."[2]

Let me remind you of some of the counterexamples:

(1) (David Christiansen): You have just taken the notorious drug LSQ, which tends to cause the person who has taken it to believe that he can fly. The drug has not yet taken effect, but you believe that you will soon believe, with 99 percent certainty that you are able to fly. But you don't now adjust your credences in the ability to fly to 99 percent. Instead, you take precautions to prevent yourself from acting on your anticipated belief.[3]

(2) (A more mundane example, from William Talbott): You've planned your evening meal, and believe with probability close to 1 that you will have spaghetti for dinner tonight, but you also believe that in a year's time you will have forgotten what you had for dinner on this particular night. Since you have spaghetti about one night out of ten, on average, and expect to remember that fact, you expect to believe (in a year's time) to degree one-tenth that you had spaghetti on this date. But it would not be reasonable for you now to lower your present credence to one-tenth.[4]

In response to such examples, the defender of the reflection principle sometimes adds the qualification that there must be no expected "cognitive impairment," which includes both expected memory failure, or apparent memory failure, and expected irrationality. It also includes cases where you expect that at the future time you might take something to be evidence that is not actually veridical evidence. But counterexamples need not involve an expected cognitive defect or memory failure. Suppose I just expect to change my mind tomorrow—to change my credences, not by adding whatever new evidence I receive to my present evidence, and updating by conditionalization, but instead by deciding that my previous beliefs were mistaken. I will then be subject to a dynamic Dutch book, but that is no reason to believe, rationally, that this won't happen. If I expect that it will or might happen, then I will now blame the fact that I am subject to the Dutch book on my future self, recognizing that my future

[2] Van Fraassen 1984, 244. [3] Christensen 1991. [4] Talbott 1991.

self will blame my current self. But it does not follow that I will be irrational then, or that I am irrational now. (Nor does it follow, without further argument, that I must be irrational either one time or the other.)

In a situation like this, one is inclined to say that the agent is not rationally responsible, at either time, for his or her beliefs at the other time.

3. The Qualifications

To avoid the counterexamples, I want to suggest that we need two different kinds of qualification to the principle, one more difficult to pin down than the other. The easy one is this: the principle holds only on the condition of perfect recall—no forgetting. We might put the condition this way: One expects to have all the evidence at $t+x$ that one had at t, plus perhaps more.

Our second, harder-to-pin-down condition aims to exclude cases of expected irrationality, cases of arational shifts in one's belief state, and cases of uncertainty about the reliability of one's future judgments about what evidence one then has. The condition, which will need considerable explanation, is that one must *endorse*, at the earlier time, one's credence state at the later time.[5]

I will try to say more about exactly what endorsement comes to, but I may not say enough to forestall a worry one might have: one might argue that the condition trivializes the principle—that in effect, when we add this qualification, we are really saying just that reflection holds except when it does not. But we will see there are still interesting problem cases to consider, even for the principle qualified in this way. And even if the qualification succeeds in making the principle invulnerable, the relevant notion of endorsement does have some interest, and I think it helps to bring out some features of the dynamic situation.

What does it mean to endorse a credence state, or a state of belief, however it is represented, in a case where the state one is endorsing (at a certain time) is different from one's own (at that time)? The general idea is that when I endorse a belief state, I am judging that it is the reasonable belief state to be in for a person who has the evidence that the person then has. In nontrivial applications of this notion, I won't know exactly what specific state is in question (since I won't know what the person's evidence is). I can still endorse it by judging that the person (me at a later time, or someone else) will assess his or her evidence in the way that I think it should be assessed. In the specific context of a representation of a belief state by a probability function, endorsement could be modeled by a judgment that there is a common prior shared by me now, and the other agent (who might be me at a different time.) The intuitive notion of endorsement that I am trying to get at is more general, and might be applied to other ways of representing a belief state, or an epistemic situation, but the general notion does presuppose that we can distinguish (1) a cluster of methods and general beliefs about the world that might be shared by agents in different belief

[5] Van Fraassen uses the word "endorse" in defending his reflection principle, suggesting that when you are a rational agent, you undertake a "commitment to follow only epistemic policies which you can endorse" (van Fraassen 1995, 26).

states from (2) more local features of a belief state—that is, from the particular evidence that each of us has. Endorsement is the judgment that I have enough in common with the person whose state is being endorsed so that the divergences in our belief states can be explained by the difference in the evidence available to us. What the reflection principle does, when it is qualified in this way, is to spell out a constraint on the relation between a belief state X and a belief state Y that is endorsed by one who is in belief state X.

The *common prior* used to model the notion of endorsement in a Bayesian context need not be some kind of a priori initial probability function (an ur-probability, to use a term that Chris Meacham and others have used for such a notion[6]). According to Alan Hajek, David Lewis used the term "Bayesian Superbabies" for logically perfect rational agents "as they begin their Bayesian odysseys." Their credence function encodes what one's credences should be before receiving any empirical evidence.[7] But I want strenuously to disassociate the picture I am trying to develop from any such notion. I have no idea what an ur-probability function might be, or where it might come from. Personally, I don't believe I had any credences at all until I got beyond the baby stage, and by the time I acquired some, they were irremediably infected by my exposure to empirical evidence. The only way I can understand a prior probability function is as one that characterizes an agent's epistemic state prior to the receipt of evidence in a particular situation, or perhaps as a function abstracted from one's credence function at a particular time (a hypothetical prior credence function that might be what one believed before receiving certain evidence that one has).

Timothy Williamson's notion of epistemic probability presupposes a measure of "intrinsic plausibility of hypotheses prior to investigation" that determines the degree to which any hypothesis is supported by the evidence one acquires, but he says little about its status, or where it comes from.[8] The label "intrinsic" suggests an a priori notion, but Williamson says that it may be context-dependent, and he does explicitly reject as "moribund" the Carnapian project of constructing a notion of logical probability. But is the measure of intrinsic plausibility we are using, or ought to be using, something that might be revised in response to experience? Might it be that it is a *contingent* fact that a certain prior probability function is the right one to have? In my view, our empirically guided conception of what the world is like develops together with the methods that we use to change that conception. Locally, we may factor out some of our beliefs and empirically motivated principles for revising beliefs that are more stable, and that are shared with peers and (normally) with our past and future selves, from others that we expect may more easily change, or be in dispute. That rough and ready, and highly context-dependent distinction is all we need to motivate the general notion of endorsement, and the more specific notion of a common prior (between a pair or group of credence states) that we use to model the notion of endorsement in the context of a Bayesian framework.

[6] See Meacham 2016. Meacham distinguishes a number of different interpretations of the notion of an ur-prior, but says that on any of the interpretations, an ur-prior will represent what a rational agent's credences should be before receiving any evidence.

[7] Hajek 2010. [8] Williamson 2000, 211.

4. The Generalization

Van Fraassen's reflection principle, as stated, applies to credences at one time about beliefs at a *later* time of the *same* agent, but neither the temporal order nor the identity of the agent is necessary. The two qualifications that I suggested are needed for the reflection principle to be defensible, but once these qualifications are in place, one can generalize the principle to include an agent's credences about his own credences at an earlier, as well as a later time, and to include an agent's credences about another agent's credences. It might be that one has special reasons (under normal conditions) to endorse one's own future credences, but as the counterexamples show, one does not always do so, and it is not always rational to do so. But it also seems that one sometimes has reason to endorse one's own *past* belief state or the belief state of a different agent. So, we can generalize the reflection principle in this way:

If agent **a**, at time t, endorses the credences of agent **b**, at time t^*, and if **a** believes that **b** has all the evidence available to her, plus perhaps more, then

$$P_{a,t}(\phi|P_{b,t^*}(\phi) = r) = r.$$

Cases where the agent **b** is the same as **a**, but the time t^* is in the past of t, rather than the future, will include cases, which are not uncommon, where the agent has forgotten the evidence on which a past credence in a particular hypothesis was based, but remembers what the credence value was. Assuming the agent endorses the credence state of his past self, this will be enough for those past credences to constrain his present ones. In cases of deference to experts, one may endorse, and it may be reasonable to endorse, the credences of a different agent (at least concerning a limited subject matter). In such a case, knowledge of the credences of the expert will constrain one's own credences in specific ways, even if one knows nothing about the evidence on which the expert bases her credences. So, I think the generalized principle, even with its qualifications, can do some work, and even in cases where one does not endorse the belief state of another, the notion of endorsement may have a role to play. It might, in a discussion with another agent with whom one has some disagreements, be a defeasible presumption that the explanation for the disagreements is to be found in differences in the evidence available to the different agents, and that the disagreements might therefore be resolved by the sharing of information. The tentative hypothesis mutually accepted in such a case, is that each is in a position to endorse the beliefs of the other.

5. Luminosity

Still, to defend even our heavily qualified principle we need to make some controversial assumptions. Jonathan Weisberg, in his incisive and illuminating critique of reflection principles emphasizes that the defense of reflection implicitly makes an assumption of *luminosity*.[9] Weisberg uses this Williamsonian term for the following assumption:

[9] Weisberg 2007.

For any hypothesis H, and time t, if $P_t(H) = r$, then $P_t(P_t(H) = r) = 1$.

Without assuming this kind of luminosity, there will be counterexamples even to our highly qualified reflection principle. I will go through an example to show the connection between luminosity and reflection, but first I will look briefly at a general problem that arises for both the luminosity principle, the reflection principle, and for any consideration of an agent's attitudes toward her attitudes (at the same or different times).

The general problem is this: Any iterated propositional attitude claim (a claim about *a*'s knowledge or credence at time *t* in a proposition *about a*'s knowledge or credence at *t*, or some other time) will involve a reference, in the statement of the content of the attribution, to the agent and a time, and it might be that the agent *a* does not realize that she is agent *a*, or does not realize at a time *t* that the time is *t*. In this case, it might be that *a* knows at time *t* that ϕ, and is in a sense fully aware and knowledgeable about her knowledge, but still not be in a position to know the truth of the proposition that *a* knows, at time *t*, that ϕ. A familiar example: Poor Lingens, the amnesiac, might know that it is raining (in Cambridge, where he is) and he might know that he knows it, but still not be sure that *Lingens* knows it, since he is not sure he is Lingens. Suppose he has been doing research into his identity, and is now .7 sure that he is Lingens. But he still can't rule out for certain a contrary hypothesis, according to which he is Lauben, and Lingens is off in California, oblivious to the weather here in Cambridge. So while $P_{L,t}$ (it is raining in Cambridge at t) = 1, $P_{L,t}$ ($P_{L,t}$ (it is raining in Cambridge at t) = 1) is only .7, and so the luminosity condition is violated.

The notorious Sleeping Beauty puzzle is a case where an agent has beliefs about what she will believe at a different time, while recognizing that at that time she will not know what time it is, and this gives rise to apparent violations of the reflection principle, at least on what I think is the most plausible solution to the puzzle.

Let me remind you of the story, and the puzzle[10]: Sleeping Beauty (SB) is to be subjected to the following experiment: She will be put to sleep on Sunday night, and woken up on Monday morning, and then put to sleep again. A fair coin will then be flipped, and if (but only if) the coin lands tails she will be woken up again on Tuesday morning, after having been given an amnesia-inducing drug that will ensure that she will, on Tuesday, have no memory of the Monday waking. She knows all of this on Sunday, and will continue to remember these general facts about the experiment as the days go by. The question is, what should she believe, when she is awakened on Monday, about whether the coin did or will land heads? Some (the "thirders") argue that she should believe, to degree 1/3 that the result of the coin toss is heads, since there are three equally likely possibilities: that it is Monday, and the coin will land heads, that it is Monday and the coin landed tails, and that it is Tuesday and the coin landed tails. Others (the "halfers") argue that since SB knew in advanced that she would be woken up at least once, she does not learn anything new upon being woken up, and so should retain her prior credence of ½ that the result of the flip will result in

[10] Elga 2000.

heads. I have argued for the thirder solution.[11] Let us suppose that SB accepts this solution, and expects to revise her beliefs accordingly.

Sleeping Beauty believes, on Sunday, that on Monday she will not lose any of the knowledge she had on Sunday, and that her credence in the proposition that the coin will land heads will change then from 1/2 to 1/3. But it would obviously be wrong for her to revise her credence in advance, on Sunday to 1/3 heads, as our reflection principle seems to recommend. Some take this to be a reason to go with the halfer solution. Others take it as a counterexample to the reflection principle.

It is true that the following formula will be true (where "Pr_S" and "Pr_M" are SB's credences on Sunday and Monday, respectively, and "H" is the proposition that the coin lands heads):

$$Pr_S(Pr_M(H) = 1/3)).$$

Since she accepts the thirder solution, she will *endorse*, on Sunday her credences on Monday, so reflection may seem to apply, but the problem is that she knows (on Sunday, and throughout) that she will *not* know, on Monday, that it is Monday. Timothy Williamson introduces in another context (a diagnosis of a classic skeptical argument[12]) some useful terminology for distinguishing two ways of identifying an epistemic situation that one may be in: first, one can identify the situation *indexically* as *this* situation—the one I am in. Second, one can identify it *descriptively* as one meeting certain descriptive conditions. Suppose a person is in an epistemic situation fitting a certain description. We can ask whether the person in that situation is able to distinguish it (identified indexically) from some different epistemic situation (given descriptively.) In this case, SB, on Monday, cannot distinguish *this* epistemic situation (as she might put it then) from the situation she will or might be in on Tuesday (given by the description, "the situation SB will or might be in on Tuesday").

Is this a genuine counterexample to our reflection principle (requiring further qualification), or can it be explained away as a merely apparent counterexample? I think the counterexample can be explained away, since the move from the Sunday to the Monday epistemic situation involves both a gain and a loss of information. The general point for present purposes is that any comparative claims about the informational content of different epistemic situations will depend on the way the possibilities used to characterize those situations are calibrated. To ascribe attitudes, what the agent does is to represent, from her own perspective, a range of possible states of the world, and considers how the agent from the other perspective distinguishes between those same possibilities. To compare her own epistemic situation with a different one—to see whether information is gained or lost—the agent needs to calibrate the possibilities as she distinguishes them from her present perspective with the possibilities as distinguished from the other perspective. Even in situations where self-locating information is not obviously involved, it is not always clear how to do this, since we represent the possibilities that epistemic situations distinguish between in terms of what we know and believe about our own epistemic situation.

[11] See Stalnaker 2011b.

[12] I will discuss the skeptical argument and Williamson's diagnosis of it in the last section of this chapter.

We fix the reference of certain terms (names, predicates, demonstratives, etc.) to things and properties that we find in the world, and we individuate alternative possibilities in terms of those things and properties. The problem of calibration is a general one, but it is most salient in cases involving one agent's attitudes to the self-locating attitudes of another (or of oneself at a different time).

I am going to set this problem aside by restricting attention to cases where the identification of the subject and time of the different credence states being considered is unproblematic, but the luminosity principle is controversial even with this restriction. Let me show why it is necessary for a defense of reflection with a familiar puzzle, one with a standard and uncontroversial solution: the Monty Hall problem.

A quick reminder of the problem: There is a prize behind one of three doors, with an equal chance for each. Alice chooses one (say door 1), and then Monty will open one of the two remaining doors to show her that the prize is not behind that one (say that door 2 is then shown to have no prize behind it). You are then offered the opportunity (for a small fee) to switch your choice to door 3. Should you do it? The puzzle is generated by the following fallacious argument: if you conditionalize on your evidence—that the prize is not behind door 2—your new credence that the prize is behind your chosen door goes to 1/2, as does your posterior credence that it is behind door 3. So, there is no benefit in switching. This argument also seems to provide a counterexample to reflection. You knew before the door was opened that you will learn either that the prize is not behind door 2, or that it is not behind door 3, and the conditional probability that the prize is behind door 1, conditional on it not being behind door 2, or conditional on it not being behind door 3, is 1/2. So, you know before receiving the evidence that whatever the evidence is, your credence will go from 1/3 to 1/2.

The fallacy is that the *total* evidence that you will receive is not that the prize is not behind door 2 (or not behind door 3), but rather *that you learn* that it is not behind door 2 (or that you learn that it is not behind door 3). When you receive evidence E, you normally also learn that you receive evidence E, and for this reason you know that the possible items of total evidence that you will receive form a partition. The proposition that the prize is not behind door 2 is compatible with the proposition that it is not behind door 3, but *that you learn* that it is not behind door 2 (under the known conditions of the game) is incompatible with the proposition *that you learn* that it is not behind door 3. Posterior credence that the prize is behind door 1 on the total evidence will remain the same (1/3) as the prior credence, whichever piece of evidence you receive, so reflection is not violated (and you should definitely switch your choice). But this standard diagnosis presupposes Weisberg's luminosity assumption. Suppose that Alice is a Williamsonian who believes that knowing that one knows that p requires a higher threshold of evidence *about* p than is required for simply knowing that p. Now suppose Alice has chosen door 1, and knows that Monty will open one of the two other doors (either 2 or 3), and she knows that she will know which of the two doors it was that was opened, but the evidence about which of the two doors it is will be just strong enough for knowledge of that fact, and not strong enough for knowledge that she knows that fact. So (we are supposing) Alice knows, in advance, that her *total* evidence will be either "the prize is not behind door 2" or "the prize is not behind door 3." Given these assumptions, it will follow that in the prior situation, she knows that her posterior credence (after the door is opened) will go up to 1/2, but this is not a reason to raise it now.

But for this to be a counterexample to the reflection principle, we need to be sure that the "no forgetting" assumption is satisfied, which means that Alice's prior knowledge that she would come to know not-B or else come to know not-C must be preserved. So, the anticipated posterior situation will be one in which Alice knows that either she knows that the prize is not behind door 2, or she knows that the prize is not behind door 3, and that she knows only one of these two things. But she does not know which of the two she knows. The posterior situation, assuming things go as planned and anticipated, will be somewhat bizarre, as illustrated by the following dialogue, where Alice, the player in the Monty Hall game, strives to be as informative as she can, but is careful to conform to the knowledge norm of assertion:

MONTY: Is the prize behind door 2?
ALICE: No, definitely not.
MONTY: Is it behind door 3?
ALICE: I can't say.
MONTY: So, you don't know that it's not behind door 3?
ALICE: Sorry, I can't say about that either. I do know of either door 2 or 3 that it is not behind that door, but I can't say which it is that I know. I can also tell you that I know *exactly* one of those two things—that it is not behind door 2, or that it is not behind door 3.
MONTY: But you already told me that it is definitely not behind door 2, right? So, don't you know that?
ALICE: I didn't say that I *know* it—I just said that it's true.

On hearing this dialogue, I would conclude that if Alice is sincere, and not just playing mind games with Monty, then she suffers from a serious pathology—a kind of self-blindness that one might expect to find in one of Oliver Sacks's cases. Williamson's views about the failure of the KK principle seems to suggest that rational and responsible knowers could find themselves in situations where responses like Alice's are rationally required.[13] If so, then I think there must be something wrong with the view. Williamson's "margin of error" arguments against the KK thesis are important and challenging; they deserve a response, and I have tried to give one elsewhere,[14] but for now I am just going to accept the kind of transparency assumptions that suffice for Weisberg's luminosity principle (at least for the case where the agent and time are identified in a way that is accessible to the agent at the time.)

It should be noted that even if one were to buy the KK thesis, no one should accept the inference from not-K to K not-K. Knowers are not infallible and even a fully rational agent might mistakenly believe that she knows something when she in fact does not. If this is possible, as it surely is, then there will be cases where a knowledge

[13] However, there are principles weaker than KK that Williamson might accept that would rule out the bizarre epistemic situation we are attributing to Alice. For example, one might, without endorsing KK, hold that if a rational agent knows that he either knows that ϕ or knows that ψ, but not both, then she must be in a position to know which it is.

[14] Stalnaker 2015.

claim is false, but the knower does not know that it is. Recognizing this fact does not affect Weisberg's luminosity principle, which is about credence, but does it threaten the viability of the qualified reflection principle? It is true that the defense of the reflection principle assumes that the candidates for posterior total evidence will form a partition. and it is also true that to know what your total evidence is, you must not only know that you have the evidence, but also that it is all of your evidence. If you take yourself to know something that in fact you do not know, then you will be mistaken about what your total evidence is, and as a result it could happen (even assuming the luminosity condition) that the candidates for total evidence do not form a partition, relative to a prior knowledge state. This will not, however, be a problem for our qualified reflection principle for the following reason: It is part of the idea of endorsement that an agent **a** (at time *t*) endorses agent **b**'s credences (at time *t'*) only if **a** presupposes that **b** really has the evidence that he then takes himself to have. So, whether or not **b** is *in fact* mistaken about what his total evidence is, **a** will endorse **b**'s expected credences only if **a** *believes* that **b**'s total evidence coincides with what **b** takes his total evidence to be. This will be enough to ensure that, in cases where the reflection principle apples, **a** will take the alternative candidates for **b**'s total evidence to form a partition, as I will argue below in the context of an example.

6. An Example

As I said at the start, my main concern is not to defend a version of the reflection principle, but to reflect more generally about the epistemic role of an agent's knowledge, beliefs, or credences at one time about her own knowledge, beliefs, or credences at another time, or about those of a different agent. That is, we are concerned with cases where an agent, from a certain epistemic perspective, ascribes attitudes to an agent who has a different perspective. The reflection principle is one constraint that will play a role, in some cases, in an agent's reasoning about what to believe, and how to revise his or her beliefs. One's judgments about what others (or oneself at other times) believe may constrain how one's credences should change in response to given evidence, but may also play a role in determining what should count as evidence. To bring out some of the dialectic of epistemic reasoning about relevance of beliefs about different epistemic situations, I will conclude by looking in detail at an example—a toy model of an epistemic situation—that raises prima facie problems for our reflection principle.

In his discussion of evidential probability, Williamson gives a simple model of an epistemic situation to illustrate what he calls "a puzzling phenomenon," and that he also takes to be a counterexample to a reflection principle like the one I have been defending. The model is given in the classic Hintikka-style semantic framework for epistemic logic, but with the addition of probability measures on the spaces of epistemic possibilities. The probabilities are given by a single prior probability distribution over all the possible worlds. The probability function representing an agent's epistemic probability in a given possible world is the restriction of this prior to the set of worlds that are epistemic alternatives to that world. Williamson's specific model represents just three candidates for epistemically possible worlds, and a uniform prior probability distribution on them (1/3 for each). The epistemic

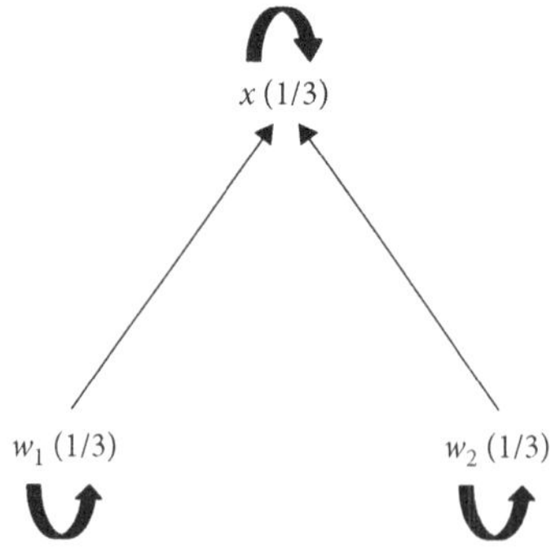

Figure 5.1

accessibility relation determining what the agent knows in each of the possible worlds is shown in Figure 5.1:[15]

Here is the story that goes with it:

> A simple creature monitors the ambient temperature by means of two detectors. When it is not cold, the first detector is activated and causes the information that it is not cold to be stored; otherwise the first detector is inactive. When it is not hot, the second detector is activated and causes the information that it is not hot to be stored; otherwise the second detector is inactive. The relevant three (partial) worlds are w_1 (it is hot), w_2 (it is cold), and x (it is neither hot nor cold).[16]

The puzzle for the reflection principle, as Williamson presents it, is this: Let p be the proposition that the creature is in world w_1 or w_2. The prior epistemic probability of p is 2/3. In world x, the evidence rules out both w_1 and w_2, so the posterior probability of p is 0. In w_1 the evidence rules out w_2, so the posterior probability of p is ½. In w_2, the evidence rules out w_1, so the posterior probability of p is ½. So, the creature knows before receiving any evidence that the posterior probability will drop at least to ½. Since it knows that it will know more after receiving evidence, and knows that the total evidence, whatever it is, will be evidence against p (evidence that lowers the probability of p from the prior), if we assume the reflection principle, why should we not assume that the creature, if rational, will lower its probability in p in advance of getting the evidence?

There are two reasons why this argument is not directly relevant to the particular reflection principle under discussion. The first is that Williamson's epistemic probability, as he emphasizes, is not credence (degree of belief), and not even idealized credence. It is a measure of epistemic support by the evidence. Van Fraassen's reflection principle, and our generalization of it, are about rational credence, but we can assume, first, that in a case where what the agent takes herself to know coincides with what she knows, her credence, if she is rational, will coincide with her evidential probability, and more generally, that even in cases where she is mistaken about her total evidence, her credence will be determined by what she takes her total

[15] Williamson 2000, 231. [16] Williamson 2000, 233–34.

evidence to be. This fits with Williamson's "knowledge first" picture, according to which "to believe *p* is to treat *p* as if one knew *p*—that is, to treat *p* in ways similar to the ways in which subjects treat propositions which they know."[17] Extending this point to credence, we assume that credence is probability conditional on the agent's total beliefs.[18]

The second reason that the example is not directly relevant to van Fraassen's reflection principle is that his principle is about an agent's credences at one time about its credences at a different time. Williamson's model is not explicitly dynamic in this way. What it represents is the creature's knowledge and epistemic probabilities at a single time within each possible world, a time at which it has the evidence that determines the posterior probability. Williamson acknowledges that his model is not dynamic, and notes that the prior need not represent what the agent believes at some prior time, but says that the example "is more vivid" if we think of the prior probability function as the agent's evidential probabilities at an earlier time. But as we will see, it makes a difference to the assessment of the reflection principle whether one thinks of the posterior evidential probability function as the result of updating from a prior epistemic state. I will follow Williamson's more vivid presentation, assuming that at each of our worlds x, w_1, and w_2, there are two successive times, t_0 and t_1, within each world. At each world, the creature is in the prior knowledge state at time t_0 (assigning equal 1/3 credence to each world). At time t_1 the creature is in the different posterior knowledge states that are given in the diagram.

Williamson's stipulations about the case concern what the creature *knows* at the posterior time, and not what it *believes* (what it takes itself to know), but given the luminosity assumption that we are making, we can determine the beliefs and credences from the knowledge states.

The luminosity principle, recall, is this, with the probability function, P_t, representing credence (at time t), not epistemic probability:

For any hypothesis H, and time t, if $P_t(H) = r$, then $P_t(P_t(H) = r) = 1$.

Williamson is, of course, not sympathetic to this luminosity assumption, but his model is consistent with it (and as he notes, this particular model satisfies the KK principle). So, it is worth spelling out where the argument against our reflection principle goes wrong when we make this assumption.

The assumption implies that agents know what they fully believe, and what they don't fully believe. Since world x is compatible with what S knows (at t_1) in world w_1, this implies that in w_1, and in w_2, the creature has the same beliefs (and so the same credences) that it has in x. But in x, the creature knows it is in x, and so believes that it is in x. Therefore, in w_1 (and in w_2) the creature believes (with credence 1) that it is in world x. That is, in both of those worlds, the creature mistakenly takes itself to know

[17] Ibid., 46–7.

[18] Williamson considers and rejects the suggestion that one should identify evidence with belief, rather than with knowledge, but this is not what I am suggesting here. I follow Williamson in assuming that the agent who is mistaken about what she knows is mistaken about what her evidence is, but she will still form her credences on the basis of what she takes her evidence to be, which is what she (fully) believes.

that it is in world *x*—that it is neither hot nor cold. (This interpretation does not fit with the story Williamson tells about the temperature monitors, but it does fit with the model.)

Now go back to the prior time (in any of the worlds) and consider the creature reflecting on its epistemic possibilities. It knows that there is a 2/3 chance that it will take itself to know something that it does not in fact know, and so it will therefore not endorse the credences that it will then have. This is thus not a violation of the qualified reflection principle, which applies only when the agent does endorse the expected credences.

This is enough to avoid the counterexample, but it is not the end of the story. An agent's reflection, at one time, about what it will believe at a later time can also give it reason to draw conclusions about what it will *know* at that later time—what its evidence will be. Imagine the creature, reflecting at t_0, about what evidence it will have at t_1. Williamson's model just stipulates, by its specification of the accessibility relation, what is known at that time in each world, but let's consider what the story might be about what happens to give the creature the beliefs that it has about what it knows. Since the creature believes (at t_1, in all three possible worlds) that it is in world *x*, we should assume that in all three worlds it will observe (or seem to observe) at time t_1 that *both* sensors are activated, and on this basis will conclude, in all three worlds, that it is in world *x*. The sensors will both be working normally in world *x*, but one of the sensors will be defective in world w_1, and the other will be defective in w_2, so in each of those two possible worlds it will not know what it takes itself to know. Now the creature, at the prior time t_0, knows all this, in all three worlds: it knows that there is a 2/3 chance that it will take itself to know something that it does not in fact know. Perhaps it will have forgotten, at t_1, what it knew at t_0 (in which case we have a second reason why the reflection principle does not apply), but suppose the agent expects correctly to remember, at t_1 what it knew at t_0. Wouldn't this be sufficient reason for the creature to conclude that the messages do not give knowledge after all, even in the world *x* where they are both true? And if, when time t_1 comes, the creature still remembers what he knew at t_0, shouldn't it, if rational, conclude that it hasn't really learned anything after all?

But (the Williamsonian might protest) isn't this response just a version of the standard skeptical argument that mistakenly assumes symmetry in the evidence available in the "good" case where things are normal, and in the "bad" case, represented by a skeptical scenario? Williamson's three-world model, interpreted as I have interpreted it, is just a slightly more complicated version of the simple good case/bad case model of a skeptical argument, with *x* being the good case, and w_1 and w_2 each being half-bad cases. Williamson's characterization of the skeptical argument in the simple two-world case was this: "Skeptics and anti-skeptics agree that in the bad case, one cannot discriminate the bad case, presented indexically, from the good case, presented descriptively as 'the good case'."[19] That is, it is agreed that an agent who is actually in the bad case does not have any evidence that gives him reason to conclude that he is not in the good case. But the skeptic mistakenly infers from this

[19] Williamson 2000, 168.

that an agent who is actually in the good case does not have evidence that discriminates the good case, presented as "the case I am in" from the bad case, presented descriptively as "the bad case." A rational agent can be mistaken about what he knows, and so about what evidence he has, and that is enough to defeat the symmetry assumption that the skeptic's argument requires. The same point applies to the three-world model: we cannot infer from the fact that the creature cannot distinguish, in world w_1 (or w_2) the world it is in from world x, that it therefore cannot discriminate, when it is in world x, the world it is in from worlds w_1 and w_2.

I agree with Williamson's diagnosis of the skeptical argument, and I agree that the point applies to the three-world model on a static interpretation where the prior probability function is just a way of representing the individual posterior evidential probability functions. But suppose one fills out the simple good case/bad case story by adding a dynamic dimension that provides a prior epistemic probability function assigning equal probability to the two cases. Yesterday you were told that on the following day you would walk into the room and observe what will appear to be a book on the table. While there is a 50/50 chance that you will see exactly what you seem to see, a book on the table (the good case), there is also a 50/50 chance that what you seem to see will be the result of an extremely effective trick with mirrors, and in this case (the bad case) there will in fact be no book on the table. Today, you walk into the room and see a book on the table (it's the good case). What should you take yourself to know, and what should your degrees of belief be about whether you are in the good case or the bad case? You remember today, let us suppose, what you were told yesterday, and you still believe it. In particular, you continue to believe that the trick will or would be effective in the sense that if you were in the bad case you would not be able to discriminate the situation you are in (the bad case, identified indexically) from the good case (identified descriptively, as the good case). Isn't it clear, about *this* version of the story, that anti-skeptics as well as skeptics should say that you ought to have credence ½, even when you are in fact in the good case, that you are in the good case? Isn't the story about how we get into the epistemic situation we are in relevant (at least when we know the story) to determining what the epistemic situation is—what we in fact know in the situation?

As we noted above, Williamson adopts the dynamic story just for vividness. "'Know in advance'," he says, "could just mean 'know a priori'."[20] But where do these a priori probability judgments come from and what do they mean? Forget for a moment the dynamic version of the story. Assume that you just see the book, and take yourself (reasonably) to be in the good case. But (being a philosopher) you contemplate what you correctly take to be a counterfactual possibility—that you are in the bad case—and ask how, on this hypothesis, your *epistemic* probability should be divided between the good case and the bad case. You know what your *credence* should be: you should believe (with probability one) that you are in the good case. This follows from the assumption that your credence should be what you take your epistemic probability to be. I have no idea what your *epistemic* probability is, or should be in the bad case, since I have no idea what role it is playing in assessing your

[20] Ibid., 231, footnote 11.

epistemic situation. One might invoke here the superbaby myth. Suppose you were a fully rational creature, at the very beginning of your Bayesian odyssey, not yet having acquired *any* empirical evidence. What should your prior credences be (in a bad case) about whether you are in a good case or a bad case conditional on getting apparent evidence that you are in a good case? (Apologies for the convoluted character of this question, but it is a question our logically omniscient superbaby should have no problem with.) Perhaps the dynamic story is not just adding vividness, but is also engendering an illusion that we know what we are supposing when we suppose that our judgments about degrees of evidential support are based on a prior probability function to which we have a priori access.

But while I think we should be skeptical about a prior probability function that is a priori knowable, I do think (as suggested earlier) that it is appropriate to think of a prior, not necessarily as an actual credence function at some earlier time, but as a representation of the more stable features of one's epistemic situation, a representation that abstracts away from the evidence one just happens to have received. This hypothetical prior perhaps represents the part of one's epistemic situation that one presumes is shared with one's peers who have different local experience, and so different evidence. I agree that it is heuristically useful (more vivid) to think of such an epistemic state as representing what one's credences were at a hypothetical earlier time, before having received the information one is claiming to have received. In the simple good case/bad case scenario, the agent can reflect on what he knew and was in a position to learn before seeing the book on the table. "What is my evidence that there is a book on the table?" he asks himself. He might envision contrasting situations in which he enters the room and sees nothing on the table, or just a vase of flowers, or a ceramic pig. Even if those alternatives do not occur to him before entering the room, they are alternatives he did not rule out in advance, but that were ruled out by the evidence he received when he saw the book on the table. But the skeptical scenario was not one of the possibilities that got any positive credence at all in this hypothetical prior state of belief. He can reflect on the abstract possibility that he is in the bad case, as Williamson's protagonist does, but the bad case is not thereby one of the possibilities that he takes to be open from the prior epistemic perspective, but ruled out upon receiving the evidence he receives.

6

Rational Reflection and the Notorious Unmarked Clock

1. Introduction

Gareth Evans famously observed that when you are asked (or ask yourself) what you believe, for example about whether there will be a third world war, you do not look inward, examining your state of mind, but rather look to the world, answering the question about your state of mind by considering the question whether there will be a third world war. Similarly, it seems, if asked what you *ought* to believe about the question, you do the same thing. Your thoughts, as you deliberate about what you should believe will normally be pretty much the same as when the question is about what you do believe. But we sometimes take a spectator view of our beliefs, reflecting critically on the question whether what we *do* find ourselves believing about some matter is what we *should* believe. If we are rational, our actual beliefs will be influenced by the result of this reflection. If we decide that the evidence does not justify a credence in some proposition that is as high as the one we find ourselves with, the reflection will lead us to lower it. Of course we may not be sure how much support the evidence we have gives to a particular proposition, but one might think that rationality at least requires that, on the supposition that the evidence supports proposition **P** to a certain degree, this is one's actual credence in **P**. David Christensen considers a principle, which he labels "*Rational Reflection*," that aims to capture this idea, and uses a puzzling example (the unmarked clock) to raise a problem for it.[1] Adam Elga considers a slightly different version of the Rational Reflection principle, and uses the same puzzling example to criticize it.[2] This chapter is my response to the problem posed by Christenson and Elga, and to their arguments about it. My response is complicated: On the one hand, I will argue that the clock example, properly understood, does not raise a problem for the Rational Reflection principles, and that the puzzle can be dissolved without questioning them. But on the other hand, I think that Elga's discussion reveals problems for Rational Reflection that are independent of the clock puzzle. The problems, I will suggest, derive from a lack of clarity in the very idea of an ideally or maximally rational credence function (for an agent with certain evidence) to which he should defer.

Here is my plan. In section 2 I will state Christensen's Rational Reflection principle and then in section 3 describe the clock example and the problem that Christensen thinks it poses for Rational Reflection. I will then argue that the example really poses

[1] Christensen 2010. [2] Elga 2013.

no problem for the principle, though my argument will depend on some perhaps contentious assumptions about what *evidence* is, or more generally about what determines an agent's epistemic situation. But even if one were to accept my argument, that will not be the end of the story. In the remaining sections of the chapter, I will look critically at the principle and at its motivation and presuppositions. I will consider the relation between different formulations of the principle, at a different argument that Adam Elga gives against Rational Reflection, and at a weaker qualified principle that he proposes. While I am not in the end persuaded by either Christensen's or Elga's arguments against Rational Reflection, I will conclude by expressing some doubt that it is a well-defined principle.

2. Rational Reflection

Christensen begin with the sensible assumption that a rational and responsible agent who doubts that his beliefs are the ones that he ought to have in the epistemic situation he finds himself in should, as a result, change his beliefs, but he contrasts two kinds of cases. His first example is Brayden, who believes that he tends to be overly pessimistic, and so judges that the credence that he puts on some negative outcome is probably too high. This critical judgment ought, if he is rational, lead him to revise his credence by lowering it. But then there is Ava, who also doubts that her credences are just right, given her evidence, but is unsure about the direction in which they are likely to be off. Perhaps she knows that she is sometimes overly optimistic, but also sometimes prone to exaggerate the probability of a possible negative outcome. Her doubt might be reasonable, but it does not seem to motivate any particular revision of her credences. How should one think about one's beliefs in such a situation? Christensen suggests that one's doubts ought at least to influence one's *conditional* beliefs, conditional on the supposition that certain credences are the ones that one ought, given one's evidence, to have. Suppose that "Cr" is Ava's actual credence function in a certain epistemic situation, and that "P*" is a label for the credence function that she ideally ought to have, given the evidence she has in that situation. Then the principle (labeled "RatRef" for Rational Reflection) says that Ava's credence in ϕ, conditional on the proposition that $Pr^*(\phi) = r$, should be r:

$$Cr(\phi/Pr^*(\phi) = r) = r$$

Christensen's RatRef is one of a family of principles, sometimes called "deference principles," all having the same abstract form. Each states a relation between two probability functions, expressing the idea that the outer function (usually representing a credence function) should "defer" to the inner one. The agent may not know the value of the inner probability function, but the upshot is that her credences should be the expectation of the value of the inner function. In RatRef, the inner function represents, in Christensen's words, "the maximally rational response to [the agent's] epistemic situation." Two other familiar examples of principles of this kind are, first, a reflection principle proposed by Bas van Fraassen, where the inner function represents the agent's credences at a later time when the agent is better informed, and second, David Lewis's Principal Principle, where the inner function represents objective chance. In all of these cases, the label for the inner function, "P*," does not name a particular probability function, defined by its mathematical properties, but rather is a descriptive term for whatever probability function it is that plays a certain role (the ideal credence function for the agent at the relevant time, or the credence function the agent will have at a certain later time, or the objective chance function that holds in the world and at the time at which the agent has the credences represented by the outer probability function.) So, the condition, '$P^*(\phi) = r$' in the statement of the principle expresses a contingent proposition (for a given r and ϕ) that is true, relative to a world, time, and agent, just in case whatever probability function plays the stated role in that world at that time and for that agent has the value r for the proposition ϕ.

The assessment of any of these reflection principles depends, of course, on exactly what role is represented by the inner probability function. So, to understand and evaluate RatRef, we need to look critically at just what it might mean for a probability function to be the ideal or maximally rational response to an agent's epistemic situation, at what defines an agent's epistemic situation, and at the presupposition that there is always a unique function meeting this condition. In order to help get clearer about these questions I will start by looking at the example that Christensen uses to raise a problem for the principle he proposes.

3. The Unmarked Clock

The example is a variant of an example used by Timothy Williamson to make a related point.[3] It is supposed that "Chloe is looking at an unmarked clock a few feet away with just a minute hand" which is pointed at some particular minute, perhaps somewhere in the lower right quadrant.

Chloe is considering various propositions about the hand's position such as:

P19 The hand is at 19 minutes after the hour;
P20 The hand is at 20 minutes after the hour;
P21: The hand is at 21 minutes after the hour;
etc.

Ideally, what should her credences be in this situation? She should not be certain that any one of them is true—her perceptual capacities are not that good, and she knows it. Christensen suggests that she might say something like this: "If I had to bet on one position, I would bet on P21, but I wouldn't be that surprised if it was 20 or 22, and it might, for all I know, even be 19 or 23,"[4] Suppose her credences are distributed around her best guess as follows: P19: .1, P20: .2, P21: .4, P22: .2, P23: .1. These numbers are rounded off, as there will be a remote possibility of a more serious deviation between her best guess and the actual position (she might be distracted, or there could be a distortion of some kind in the perceptual conditions). But let's assume that under normal conditions, the actual position and her best guess never diverge by more than $\pm$ 2.[5]

[3] Williamson 2011. [4] Christensen 2010, 123.

[5] I've changed the numbers slightly from those Christensen uses in giving this example, but the points are the same.

We can assume that the situation will be similar if the clock hand position and the best guess were different, with the same .1, .2, .4, .2, .1 distribution around the number that she would bet on, if she had to choose one. Christensen imagines a big 60 × 60 chart that represents her powers of perceptual discrimination in this observational situation, and that determines what her credences ought to be, given the specific evidence that she has when she looks at the clock. A fragment of the chart might look like Figure 6.1:

	P9	P10	P11	P12	P13
P9	.4	.2	.1	0	0
P10	.2	.4	.2	.1	0
P11	.1	.2	.4	.2	.1
P12	0	.1	.2	.4	.2
P13	0	0	.1	.2	.4

Figure 6.1

The rows represent the alternative best guesses, and the columns represent the alternative positions of the hand. The chart might be determined by gathering statistics on the following experiment, run with many subjects who are similar to Chloe in their perceptual capacities. Each is placed a few feet away from the unmarked clock, with the minute hand in various different positions, and asked to give his or her best estimate for the exact location of the hand. The result of the experiments, let us suppose, is that the subjects are exactly right approximately 40 percent of the time, off by at most 1 approximately 80 percent of the time, and off by at most 2 almost all of the time (with the small errors on each side distributed equally). But, however one comes to know what Chloe's chart is, the chart will be *correct* just in case it describes her actual dispositions, under normal conditions, to respond to the question about her best estimate, in various different particular circumstances. We could describe the chart as determining the prior probability function that ought to govern Chloe's degrees of belief on particular occasions when she looks at the clock. The hand could be in any position for all she knows prior to looking at the clock, and she also has no prior information about what her best guess will be. So, the prior will give each of the rows, and each of the columns, a prior probability of 1/60. The numbers in the cells represent the conditional probability of a best guess, given the position of the hand, and also the conditional probability of the position of the hand, given the best guess. Christensen assumes that Chloe knows that the chart is correct.

Since the chart describes Chloe's actual perceptual capacities in the situation she is in, it seems reasonable to assume that it determines the credence function that is ideally rational for her, given her particular evidential situation, and that is what Christensen assumes. But what defines her evidential situation? Christensen takes her evidential situation to be defined by the actual location of the clock hand, so that the .4 credence in location P21 is perfectly rational "only if she's actually in the evidential situation that would be produced by the hand being at 21." The problem for the Rational Reflection principle that is raised by this example is this: Chloe is ideally rational (Christensen argues) only if her best guess is exactly right—only if, when she says, "if I had to bet on one position, I would bet on P21," the actual position of the hand is 21. But then, assuming Chloe knows the chart, she knows (and believes with probability one) that her .4 credence in P21 is ideally rational if and only if the hand points to 21. But in general, for any coherent credence function, if $Cr(X \leftrightarrow Y) = 1$, then $Cr(Z/X) = Cr(Z/Y)$, for any Z, which means (as Christensen observes) that $Cr(P21/Pr^*(P21 = .4)) = Cr(P1/P1) = 1$, and this violates RatRef.

But this characterization of Chloe's evidential situation seems to get things backwards. There is no *one* evidential situation that "would be produced by the hand being at 21." It is a feature of Chloe's perceptual capacities and limitations, as represented in the chart, that sometimes the evidential situation she is in when the clock hand is actually at 20 is exactly the same as it might be if the clock hand were at 21.

Consider how Chloe uses the chart to determine her credences, and how she *should* use it. She might say, "Based on my observation of the clock, my best guess, if I had to choose just one position, is P21, and I know from the chart that in that case, the actual position of the hand is always (under normal conditions) within the range, 21 ± 2. Furthermore, I know from the chart how likely it is to be at each of the positions in this range. For example, I know that there is only a 10 percent chance (given my considered best guess) that the hand is actually at 23." Suppose it turns out that 23 was the actual location of the hand, and Chloe is criticized for having credences that were not ideally rational. She responds, "As Nate Silver has emphasized, in his discussion of the interpretation of polling data, sometimes (in fact about 10 percent of the time) when you *correctly* claim that a certain outcome is 90 percent likely, a different outcome will obtain, and this does *not* show that the initial judgment was not exactly the right one."

If we think of the chart as representing an appropriate prior for Chloe, then the proper use of it to determine the posterior credence is to conditionalize on the best estimate. What we want is the probability of the actual position, conditional on the estimate, and not the probability of the estimate, conditional on the actual position (even though the values are, in this case, exactly the same).

A base rate fallacy is the confusion of the probability of a hypothesis, conditional on the evidence, with the probability of the evidence, conditional on the hypothesis (in statistical jargon, the *likelihood*). In routine Bayesian reasoning, the latter plays an important role in computing the former, but the numbers will be quite different when the prior probability of the evidence is very different from the prior probability of the hypothesis. If one confuses likelihood with probability of the hypothesis, one will sometimes judge the hypothesis to be much more probable than it is, on the evidence. In the clock story, the prior probability of the estimate is equal to the prior probability of the actual position of the hand, and so the two posterior probability values (of the estimate, conditional on the actual position, and of the actual position, conditional on the estimate) are the same, but it is still important to distinguish the two roles. It is clear enough, in the clock example, that the estimate is a part of the evidence—Chloe knows what her estimate is. It is also clear that the actual position of the clock is *not* part of the evidence, even if it may be relevant to determining what the evidence is. For Williamson, a proposition is evidence if and only if it is knowledge, so the total evidence that Chloe has (and so the evidence that determines her ideally rational credences) will depend on what she knows. I accept Williamson's E = K (Evidence = Knowledge) thesis, and it is uncontroversial that one can be mistaken about what one knows. So, if ideally rational credence is conditional on one's actual evidence, then I agree that one may be mistaken about the credences that one ideally ought to have. But I don't think that in the clock story, Chloe would be mistaken about what she knows, even if the actual position of the clock diverged (within the $\pm$ 2 range) from her estimate. Given her knowledge of the chart, Chloe has good reason to be confident that the actual position of the clock hand is within the $\pm$ 2 range of her estimate—in fact she can be absolutely certain, on the assumption that conditions are normal, that the actual position is within this range. So, there is no puzzle about the clock, and no conflict with the principle of Rational Reflection.

4. The Ideal, and What One Ought to Believe

In general, what one ought to believe is what is supported by one's evidence. In discussing the clock puzzle, we assumed, following Christensen, that Chloe *knew* that the chart was a correct representation of her perceptual capacities (she was informed by the epistemology oracle, who she rightly took to be a reliable source), and so the correctness of the chart was part of her evidence. Still, we should consider what to say about the situation in which the chart is correct, but not known. After all, if the truth be told, there is no epistemology oracle to provide us, magically, with information about our epistemic capacities. So, suppose that the chart is correct about her capacities, and also correct about the way she forms her credences in response to the estimates she finds herself inclined to make, but that she has never considered explicitly the question of exactly how reliable her estimates are in cases like this. So, we are supposing that the case is like the ordinary situation of an unreflective knower who takes the reliability of her senses for granted unless there is reason to think otherwise. That seems to be good enough in ordinary situations in which her senses are in fact reliable, and I think it is good enough in Chloe's situation. One might say that Chloe is implicitly presupposing that the chart is correct when she forms credences in accordance with it, and that when this presupposition, along with the presumption that perceptual conditions are normal, is correct, her credences are the ones that she ideally ought to have.

But let us also consider a more complex case. Suppose Chloe *has* reflected on the question of what chart best represents her perceptual capacities, and is uncertain. Or perhaps the epistemology oracle (to bring her back into the picture) has informed her that the correct chart is one of two, C1 or C2, but left it open which. How should Chloe (ideally) use this information to form her credences? Suppose Chloe is bold and arbitrarily opts for C1, and that she guesses right, forming her credences on the basis of the correct chart. Alternatively, we might suppose that Chloe is cautious, and splits the difference, forming her beliefs by taking a 50/50 weighted average of the results of the two charts. Which credence function is ideal, given Chloe's evidence? I think it is clear that cautious Chloe is the more rational one. If this is right, then neither of the credence functions determined by one or other of the charts represents the credences that she ought, ideally, to have.

There is a contrast between Rational Reflection and other deference principles with the same formal structure, such as van Fraassen's reflection principle, where one is deferring to the future beliefs of one's better-informed self. With that principle, your deference is based on the assumption that the person you are deferring to has (or may have) evidence that you lack. But with Rational Reflection, you are deferring to a hypothetical person who has exactly the evidence that you have. In the former case, one can recognize and accept that your credences diverge from those you are deferring to, but in the latter case, there is a built-in instability in the recognition that your credences diverge from what they ought to be. In our two-chart example, cautious Chloe is deferring to the correct chart in the way that one defers to one's future better-informed self. In both cases, since one does not know the value of the credence one is deferring to, one bases one's credence on an expectation of that hypothetical credence. But the credence based on the expectation will be different from the credence one is deferring to, and so if the expectation is the reasonable credence to have, then the credence one is deferring to cannot be the credence that one ought to have. We need to look more closely at what is meant by "the maximally or perfectly rational response" to one's epistemic situation.

Recall Christensen's example of Ava, who is equally worried that her credences may be influenced by wishful thinking, or by protective pessimism. She is like Chloe, who is unsure which of two charts correctly characterizes her perceptual capacities. Ava's learning which of the two potential biases is relevant to her particular situation would be like learning which of two charts is the correct one—whether the credences she is inclined to have are skewed up, or skewed down. Like bold Chloe, she would be wrong to opt, arbitrarily, for the hypothesis of pessimism or for the hypothesis of wishful thinking, even if she got it right. Perhaps Ava's credences are exactly as they should be, even if she is sure that either wishful thinking or protective pessimism is part of the explanation for why her credences are as they are. Perhaps they are exactly right because they are the rational response to her recognized epistemic limitations.

Ava's situation is not exactly the same as cautious Chloe's, but both are deferring to a hypothetical credence function that is better informed than the one she is in a position to have. When Ava reflects on the credences she finds herself with, she asks herself, "Why are my credences the way they are?" She might then reflect as follows: "Part of the explanation, I am sure, is that I have certain evidence, though since the evidence is complicated, perhaps part of the explanation is an irrational bias in one direction or the other. If there were a person who knew my evidence, but that I knew was free of any irrational biases, I would defer to that person. There isn't such a person, but I can say that I have a 1/3 credence that, because of wishful thinking, my credence is higher than what this counterfactual person would tell me that it should be, a 1/3 credence that it is lower because of protective pessimism, and a 1/3 credence that no biases are playing a role. So, while I am not certain what an unbiased person would tell me, I have some expectations about it that I can defer to." Given this, what *should* Ava's credences be? One view is that they should be just what they are. This is because she should *defer*, not to what her beliefs ought to be, but to what they would be if she had just the evidence that she does have, but were subject to no biases. Ava might make her two hypotheses that her credences are biased (too high because of wishful thinking, or too low because of protective pessimism) precise by specifying a corrected value, for each hypothesis, and then calculating the expectation. She would be deferring, in this case, not to what her credences ought to be, given her evidential situation, but to a certain counterfactual credence function for a hypothetical agent who is in a cognitively superior situation. Call this hypothetical agent "Ava's personal epistemic angel." Ava has at least a 2/3 credence that her credences differ from those of her angel. If the credences of Ava's angel are those that she would have if she were ideally rational, then she knows that her actual credences are probably not ideally rational. But it is not obvious that this implies that they are not the credences that, given her actual situation, she ought to have.

5. Immodesty

Adam Elga, following Christensen, took the clock story to provide a counterexample to a principle of Rational Reflection, but he also had an independent argument against the principle. So even if the clock puzzle can be dissolved, as I have argued it can, there may remain a problem for Rational Reflection. I want now to look at this argument in the light of our discussion of what is, or might be, meant by the "ideal" credence function to which the rational reflection principle says that we should defer. But before getting to that, we need to consider Elga's version of a principle of Rational Reflection, which is slightly different from Christensen's.

Christensen's RatRef, recall, is this:

$$Cr(\phi/Pr^*(\phi) = r) = r$$

"Cr" denotes the agent's credence function in a certain evidential situation, and "Pr*" is "the maximally rational response to [the agent's] epistemic situation." Elga's related principle, RR, is this:

$$Cr(\phi/P \text{ is ideal}) = P(\phi)$$

We can assume that Elga's "ideal" is equivalent to Christensen's "maximally rational response," however that elusive notion is understood.

One crucial difference between the two statements of Rational Reflection is this: the "Pr*" in our formulation of RatRef is a non-rigid term for whatever probability function it is that is playing the right role—the role of being the ideal, or maximally rational credence function for the particular evidential situation in question. Elga's "P," on the other hand, is a variable ranging over particular probability functions. So a sentence of the form "$Pr^*(\phi) = r$" will state, for a particular ϕ and r, a contingent and time and agent relative proposition, whereas "$P(\phi) = r$" (for particular values of P, ϕ, and r) will state a necessary truth or falsehood about the mathematical function that is the value of the variable, "P." "P is ideal" makes a contingent claim (given a value for P) about a particular evidential situation, whereas "P* is ideal" is true by definition, at least given the presupposition that there is something playing the role in question.

One can state the two principles with a common vocabulary this way:

RR: $Cr(\phi/Pr^* = P) = P(\phi)$
RatRef $Cr(\phi/\exists P(Pr^* = P \wedge P(\phi) = r) = r$

Given the presupposition that there is a unique ideal function, one can show that RatRef follows from RR, but the latter is slightly stronger.[6]

Now the problem that Elga raises for RR is this: first he argues that we should accept a principle labeled MODESTY: "In some possible situations, it is rational to be uncertain about what degrees of belief it is rational for one to have. Furthermore, it can be rational to have a positive degree of belief that one is in such a situation." Second, it is argued that RR implies that MODESTY is false. Therefore, we must reject RR.

MODESTY is defended with the following example (also discussed by Christensen):

> HYPOXIA:
>
> Bill the perfectly rational airline pilot gets a credible warning from ground control:
>
> Bill, there's a 99% chance that in a minute your air will have reduced oxygen levels. If it does, you will suffer from hypoxia (oxygen deprivation) which causes hard-to-detect minor cognitive impairment. In particular, your degrees of belief will be slightly irrational. But watch out—if this happens, everything will seem fine. In fact, pilots suffering from hypoxia often insist that their reasoning is perfect—partly due to impairment caused by hypoxia!
>
> A few minutes later, ground control notices that Bill got lucky—his air stayed normal. They call Bill to tell him. Right before Bill receives the call, should he be uncertain whether his degrees of belief are perfectly rational?

Of course, Bill's credences (if he is rational)—the ones about which we are asking, "Should he be uncertain whether they are rational?"—were probably themselves influenced by the initial report from ground control. Perhaps before getting the report, he was almost certain that he had enough fuel to get to the next airport, while after the report he becomes less sure. But even after adjusting his credences in response to the warning, he will still remain uncertain that he has adjusted them in the right way, and he *ought* to remain uncertain, and so be modest in the sense defined.

The second move in the argument is the following proof: By RR, we know that $Cr(\phi/P^* = P) = P(\phi)$ for any ϕ. So, we can substitute "$P^* = P$" itself for ϕ, yielding

$$Cr(P^* = P/P^* = P) = P(P^* = P)$$

So, it follows that for any candidate P for P^* (which means for any P such that $Cr(P^* = P)>0$), $P(P^* = P) = 1$. So, it follows from RR (Elga claims) that the agent with the ideal credence function knows that she is ideal, and so she violates modesty.

Now, if we follow the suggestion sketched above that what P^* represents is the hypothetical credences of a counterfactual ideal agent who is or might be distinct from the actual agent, then Elga's argument does not work. What the HYPOXIA story showed was that *Bill* should be modest, not that his epistemic angel should be.

[6] I say *slightly* stronger, since the abstract countermodels that show that RatRef can be satisfied while RR is not are contrived and fragile. See the appendix to this chapter for a countermodel.

What the proof showed is that (assuming RR) Bill's personal angel should be certain that *she* knows that her credences are the right ones for *Bill* to have, given Bill's evidence. (I will assume that Bill's angel is female, so that it will be clear who the pronouns are referring to.) We can conclude that Bill is immodest only if Cr(P*=Cr)=1, and this does not follow.

One might be tempted to argue as follows: Bill does not know what his angel's credences are, but if we accept RR, then he does know, according to Elga's argument, that his angel knows that her credences are free of all bias, and perfectly rational. So, if he defers to the angel, shouldn't he infer that he himself is free of all bias, and perfectly rational? No, he should not, since the angel's beliefs about her own ideal rationality are indexical beliefs. If Bill defers to the angel on this matter, he will infer that the angel is free of bias and perfectly rational, and not that he himself is. (Compare the reflection principle that defers to one's better-informed future self. I know that tomorrow I will know the answer to question Q, but deference to tomorrow's beliefs does not tell me to infer now that I know the answer to question Q.)

Our story assumes that we can separate what Bill, or Ava, or Chloe, ought to believe from what he or she would believe if ideally rational. But suppose instead we interpret "P*" so that our agent is rational only if his credences coincide with the ideally or maximally rational credences of an agent with exactly the evidence that the real agent has. Then if "Cr" represents Bill's credences, and P* represents the ideal credences for Bill's evidential situation, the assumption that Bill is rational is the assumption that Cr = P*. Given this assumption, we can get Elga's conclusion that if Bill is rational, and modest, then his credences do not conform to RR.

But there is reason, independent of considerations about Rational Reflection, to distinguish rationality from ideal or maximal rationality, where the latter requires conforming to all reasonable epistemic norms. It seems reasonable to assume that the credences of the ideal epistemic agent are the ones he ought to have, conditional on his evidence, and also that a rational person might be mistaken about what his evidence is. Suppose we follow Williamson in assuming that you know that ϕ if and only if ϕ is evidence that you have. Anyone who takes himself to know something that he does not in fact know will then be mistaken about his evidence, and will have credences that violate an epistemic norm. So, if we identify rationality with ideal or maximal rationality, and assume with Williamson that knowledge is a norm for (full) belief, then anyone with a false belief will fail to be rational.

Perhaps rationality—even full rationality—requires only that one base one's credences on what one takes oneself to know. Even if to have a false full belief is to violate an epistemic norm, perhaps it does not count as a deviation from rationality. But if we assume that anyone whose credences are distorted in any way by biases will be irrational, then we will have to conclude that rationality is rarely if ever found in the real world. Elga's modest Bill is an artificial case since most of us don't get testimony from a reliable source that we are 99 percent likely to be suffering from a mild cognitive impairment, but the hypoxia story can be taken as a model of the human epistemic condition. We are all in an evidential situation that is complicated and somewhat opaque, and none of us should be confident that our credences are uninfected by any biases. It might be reasonable for any of us to be 99 percent confident, as Bill is, that we are not in a position to get all of our degrees of belief exactly right. While it may not be at all implausible to think that we all deviate from ideal rationality, it does not seem appropriate to describe this kind of modesty as 99 percent confidence that one is irrational.

The issue is not just terminology: we want to ask how a rational agent should deal with a situation in which he strongly believes, or even knows, that he is not ideally rational. The agent asks, what *ought* I to believe, in such a situation—how ought I to cope with my expected epistemic impairment? Anyone trying to reason in such a situation needs to distinguish the question about what his own credences should be from the question what they would be if they were ideal, even if he uses his expectations about the answer to the latter question to decide on an answer to the former question.

On the epistemic angel interpretation, what Elga's proof shows is that RR implies that Bill's epistemic angel knows that she is Bill's angel, and so knows that her credences are the ideally rational ones for a person with exactly Bill's evidence to have. Is it reasonable to assume this? We are talking here about a hypothetical agent, so we can define it as we wish, so long as it is clear what it is that determines that counterfactual agent's degrees of belief, and so long as it would be reasonable to defer to such an agent. But let us assume that we have reason to hypothesize an angel who may herself be modest. To allow for this, Elga proposed a modified principle, called "New Rational Reflection."

NRR: $Cr(\phi/P^* = P) = P(\phi/P^* = P)$.

One should defer only to the *conditional* credences of the angel, conditional on the supposition that she is the angel. I have argued that there is no need to avoid the consequence that the angel is immodest since it does not imply that Bill himself should be immodest, but in any case, the new principle is, in effect, just a way of making the ideal to which one is deferring immodest by giving the angel the information that she is the angel. That is, the modest hypothetical angel's conditional credence function (conditional on the proposition that she is the angel) is itself a credence function, and an immodest one. Deference to this immodest credence function (by RR) is the same deference to the original modest credence function by NRR.[7] Since the angel is just a hypothetical person, it is not clear how using NRR is different from keeping the unmodified principle, with the assumption that the angel has the information from the start.

[7] Here is an argument for this claim: Suppose we define, for each P that is a candidate for the ideal (each probability function P such that $Cr(P^* = P) > 0$), a probability function P^+ as follows:

$P^+(\phi) =_{df} P(\phi/P^*=P)$

Then let P^{*+} be the nonrigid label for the P^+ such that $P^* = P$. Then for each P, the biconditional $(P^* = P \leftrightarrow P^{*+} = P^+)$ has credence 1: The left to right implication is obvious. For right to left, suppose that P_i and P_j are two candidates for which $P_i^+ = P_j^+$. Then $P_i(\phi/P^* = P_i) = P_j(\phi/P^* = P_j)$ for all ϕ. So, in particular, $P_i(P^* = P_i/P^* = P_i) = P_j(P^* = P_i/P^* = P_j)$. The former is 1 so the latter must be 1 as well, which means that $i = j$. This implies that NRR for P^* is equivalent to unmodified RR for P^{*+}. That is, the following is equivalent to NRR: $Cr(\phi/P^{*+} = P^+) = P^+(\phi)$.

I have suggested that rationality is not the same as ideal, maximal, or perfect rationality, but we should be skeptical that any of these notions identifies a sharp and clear line. We can judge that Bill responded well (that is, responded rationally) to his predicament, but to ask whether he was, overall, rational or not, is not obviously to ask a well-defined question. And the idea that there is always an *ideally* rational state that one is aiming at—that this counterfactual personal epistemic angel is well defined—also seems to me problematic. The angel is supposed to be someone with exactly the same evidence as the agent for whom she is the angel, but it is not so easy to separate evidence from bias, since the biased agent lacks evidence that he might have about his biases. Compare the situations of Ava and of Chloe, in the version of her case where she is uncertain which of two charts correctly describes her perceptual capacities. Because of Chloe's uncertainty, the charts are themselves potential evidence. She reasonably bases her credences on her expectation about the correct chart, and to do so is to defer to a hypothetical better-informed person who has this information. But also in the case of Ava, who is unsure whether she is too optimistic or too pessimistic, the uncertainty might be resolved by further evidence. She defers to her personal angel because that hypothetical person has this evidence, at least implicitly. With the Rational Reflection principles, the hypothetical agent to whom one is deferring is supposed to be, not a better-informed person, but a person with exactly the evidence that the actual agent has. We don't assume that the angel has additional evidence that enables her to overcome her bias—she just has no bias to overcome. (Angels don't need moral strength to overcome temptation since they aren't tempted.) But doesn't the absence of epistemic distortion allow one to know things that one wouldn't know if those distortions were present? So, can't we understand, in general, deference to those free of epistemic distortion as deference to those who are better informed?

There is a kind of generalized modesty that says, "Perhaps I am totally off base in the epistemic policies I am using or presupposing. Perhaps I not only lack evidence, but am basing my credences on the wrong prior, or even making a mistake in presupposing this whole epistemic framework." Such generalized modesty may also take a milder form, as in the preface paradox: "Surely *some* of my beliefs are mistaken, and *some* of my credences are higher, or lower, than they should be." But there is no particular way of changing one's policies to which this skeptical acknowledgment leads. You might think that the best way to try to deal with your concerns about the limitations of your epistemic situation is to try to see them as limitations that might be resolved by further evidence, to think about what that evidence might be, and to base your beliefs on your expectation of the result of receiving that evidence. It is not clear how to defer to yourself as you ought to be, but you can defer to real or hypothetical agents who you take to be better informed.[8]

[8] Thanks to David Christensen and Adam Elga for incisive comments on an earlier draft of this paper, which stimulated what I hope are improvements. Thanks also to Kevin Dorst for discussion about higher-order evidence, and for his insightful work on these issues.

Appendix

The two formulations of the Rational Reflection principle discussed in this chapter are as follows:

Christensen's RatRef:

For all propositions ϕ and real numbers r, if $P_0(P^*(\phi) = r) > 0$, $P_0(\phi/P^*(\phi) = r) = r$.

Elga's RR:

For all probability functions P and all propositions ϕ, if $P_0(P^* = P) > 0$, $P_0(\phi/P^* = P) = P(\phi)$.

P_0 is a probability function representing the credences of an agent in a certain evidential situation. "P*" is a nonrigid designator for the probability function representing the ideal or perfectly rational credence function for the epistemic situation of the agent with credence function P_0.

There had been some debate about the relation between the two versions of this reflection principle. It has been shown (by Elga) that RR entails RatRef. Whether they are equivalent was an open question, but we now have a proof that RR is stronger—an abstract model in which RatRef is satisfied, but RR is not. Here is the argument:

A model is triple $\langle W, P_0, f\rangle$. W is a nonempty set representing a space of possible worlds; P_0 is a probability function defined on the propositions, which are all the subsets of W; f is a function from W to probability functions: for each $x \in W$, $f(x) = P_x$ is a probability function on the same space of propositions. P_0 represents the credence function of an agent in a certain epistemic situation. P_x represents the ideal credence function for that agent to have in world x. "P*" is a nonrigid designator denoting P_x in world x, so that the sentence $(P^* = P_x)$ expresses the proposition $\{y{:}P_x = P_y\}$. (We use "P_x," etc. both as names for the probability function in sentences expressing propositions, and for the probability function itself.)

Here is a model satisfying RatRef, but not RR:

$W = \{a,b,c\}$. We specify the probability functions by a triple of numbers representing the probability value assigned to a, b, and c, respectively by that function.

P_0: $\langle 1/3.1/3.1/3\rangle$
P_a; $\langle 1/2, 0, 1/2\rangle$
P_b: $\langle 1/2, 1/2, 0\rangle$
P_c: $\langle 0, 1/2, 1/2\rangle$

There are only eight propositions, so it is easy to check that RatRef is satisfied for each of them, and for each value of r. But RR fails. For example, $P_0(\{a\}/P^* = P_a) = 1$, but $P_a(\{a\}) = 1/2$.

Reflection principles or deference principles state a relation between an "outer" probability function (usually representing credence) and an "inner" probability function (specified by a contingent condition) representing possible credences to which the agent is deferring, according to the principle. Any such principle (for example principles of deference to an expert, or to one's better informed future self) can be formulated in either of the two ways, so the abstract point—that one of the formulations is stronger than the other—applies to any of these applications. While most such principles are formulated in the way that corresponds to Christensen's RatRef, I think the motivations for imposing such principles generally support the stronger formulation. The countermodel I have constructed seems to be a loophole—a contrived case where RatRef is satisfied without its usual motivation.

Here is a simple story to go with the model: There are three doors, a, b, and c, with a prize behind just one of them. Alice's credence is 1/3 for each of the three possibilities. She knows that Bert has some information about the location of the prize that she lacks: specifically, she knows that if the prize is behind door a, then Bert can rule out b, if it is behind b, he can rule out c, and if it is behind c, he can rule out a. Her credences are an expectation of Bert's credences, so they conform to RatRef, but her credences do not conform to RR. Perhaps if Bert were better informed than Alice (knew all that he knew, plus more), she would have reason to defer to him, but in this case, Bert lacks some crucial information that Alice has: namely information about the situations in which he would have the credences that he has. Her credences happen (by contrivance) to be equivalent to an expectation of his, but the normative reason for deference to constrain her credences do not apply.

7

Expressivism and Propositions

1. Introduction

Expressivism began with emotivism, a meta-ethical theory developed and defended by Charles Stevenson and A. J. Ayer.[1] The view starts with a distinction between factual (cognitive) discourse, and discourse about values. Value judgments do not make claims about what the world is like—claims that are true or false. Instead, they express attitudes and feelings, and have the role of urging others to share the attitudes expressed, and to act on them. Emotivism was sometimes described, both by its defenders and its critics, as a *subjective* theory of value, since it grounds values in the attitudes of subjects, but central to the emotivist doctrine was a distinction between judgments *about* one's attitudes and judgments that *express* one's attitudes. A subjectivist theory of value that contrasts with emotivism holds that value judgments are factual propositions about one's own values: to say that racism is wrong is to say something like, "I abhor racism." The emotivist rejects this view; "Racism is wrong" does not make a claim about the speaker's values, or about anything. Rather it *expresses* abhorrence for racism. It is more like "Boo to racism!" One salient difference between the explicit subjectivist thesis and emotivism comes out when one considers value judgments about counterfactual situations. On the subjectivist analysis, "If I were a racist, racism would not be wrong" says something like, "If I were a racist, I would not abhor racism," which could well be true. But the emotivist who finds a situation where he himself is a racist to be especially abhorrent, can say, "Racism would be wrong even if I were a racist," since this means something like, "Boo to racism, even in situations where I approve of it."

So there are two related distinctions that are central to the original expressivist theories, and that remain central to contemporary versions of expressivism: (1) between factual judgments that aim to describe the world, and value judgments that do not, but instead express a noncognitive attitude, or promote or endorse a plan of action; (2) between an assertion that one has an attitude, and a speech act that expresses an attitude. The second distinction can be applied to cognitive attitudes such as belief and knowledge as well as to attitudes such as abhorrence or approval: An ordinary factual claim such as "Trump was the Republican nominee for president" *expresses* the speaker's belief that Trump was the Republican nominee, but does not *say* that the speaker has this belief.

Contemporary expressivism, whose most prominent developer and defender has been Allan Gibbard,[2] acknowledges its roots in emotivism and noncognititivism, but

[1] See Ayer 1936 and Stevenson 1937.

[2] Gibbard 1990, 2003, and 2012. See also, Blackburn 1993.

departs from the older theories in at least three significant ways: First, it broadens the range of concepts to which expressivist analyses are applied. Expressivism began as meta-ethics, and discourse about morality remains an important application, but Gibbard takes the central normative concept to be *rationality*. Normative discourse is about what it makes sense to believe, as well as what it makes sense to feel and to do, so an expressivist about norms should be expressivist about the concepts of epistemology. Second, contemporary expressivists develop more sophisticated semantic tools for responding to the technical problems that have been thought to plague noncognitivist meta-ethical theories. While expressivism claims that normative judgments do not state facts, the sentences used to make them have the same indicative form as factual statements, and sentences with normative content combine with each other, and with factual claims. The traditional noncognitivist did not offer any systematic account of the compositional semantics of normative-factual discourse, but contemporary expressivists have given this problem a lot of attention. Third, Gibbard and others such as Hartry Field[3] and Paul Horwich[4] have linked expressivism with a more general theory of meaning that includes a minimalist theory of truth, and that, in effect, gives an expressivist account of factual as well as normative discourse.

I will consider the application of expressivism to epistemology in section 4, but in the next two sections I want to consider more general questions about the form that an expressivist theory should take. First, in section 2, I will describe Gibbard's solution to the technical problem, as developed in his early book, Gibbard 1990, or at least one way of understanding that solution. I took Gibbard's account of what he called "normative logic" to provide a spot-on diagnosis of the problems about compositional semantics for normative discourse, and a definitive solution to the problems, but as I read more of Gibbard's later work, I came to see that my interpretation of this solution was different from his, and that I have a different understanding of the commitments of expressivism. So, after explaining and defending my understanding of Gibbard's normative logic, I will criticize, in section 3, his expressivist account of truth, which I think blurs the line between expressivism about norms and the non-natural realism that Gibbard took, at least initially, to contrast with expressivism.

2. Normative Logic

Gibbard's "first rough formulation" of his expressivist analysis is this: "to call an act, belief or emotion rational is to express one's acceptance of a system of norms that permits it."[5]

This is not quite right, even as a rough formulation, since a judgment about the rationality of an action, feeling, or belief involves, not just the acceptance of a system of norms, but also factual beliefs. The norms are conditional: they say what it makes sense to do "in a wide range of actual and hypothetical circumstances," but a judgment, for example that it does not make sense to vote for Donald Trump

[3] See the papers in Field 2001. [4] Horwich 1990 and 1998. [5] Gibbard 1990, 83.

expresses the acceptance of a system of norms that require that one not vote for him *in circumstances that we take to be actual.* So, a second rough formulation might say that to call an act, belief, or emotion rational is to express one's acceptance of a system of norms that permits in in circumstances that one takes to be actual.

Gibbard identifies three problems that an expressivist analysis of this kind faces, and then proposes a unified solution to them all. The first problem is the most familiar: the so-called Frege-Geach problem.

> Suppose we have an adequate account of the states of mind expressed by such simple statements as "it makes sense for Antony to give battle." What are we to say of more complex contexts, like "Whenever Antony does anything it doesn't make sense to do, he clings to his purpose stubbornly." Sentences of indefinite complexity... get their meanings from the meanings of their elements in systematic ways.[6]

We know how to do truth-conditional compositional semantics for purely factual sentences that express beliefs, but we need to show how to do this for normative sentences, and for sentences that mix normative and factual terms, in a way that is compatible with the expressivist program.

The second problem is the problem of communication. When Cleopatra says that some action or attitude is rational, she expresses her acceptance of a system of norms that permits it, but there will normally be many systems of norms that permit it, and nothing is said or expressed about what specific system Cleopatra accepts. Accepting a (partial or complete) system of norms is a holistic state, but a particular normative judgment is only a partial characterization of the state. We need an account of the normative *content* of an individual normative claim that is common to the different states that are compatible with it.

The third problem, closely related to the problem of communication, is labeled "the problem of *normative naivete.*" Suppose we accept what Cleopatra says, on her authority, when she makes a normative claim. "What state of mind constitutes this acceptance?... What could it be... to accept a normative conclusion—to accept, in effect, a property of a combined normative system and set of beliefs—and yet fail to accept any specific normative system that together with our beliefs, has that property?"[7] Again, the source of the problem is that a combined system of norms and state of belief is a holistic state, while the content of a particular normative claim is just a partial constraint on such a combined system.

The common solution (as I think it should be interpreted) is to give a systematic account of factual-normative content that is independent of the various mental states that these contents can be used to describe. The strategy follows the precedent of truth-conditional semantics for purely factual sentences, so let me start with an exposition of a version of that strategy that sets up the generalization to the content of combined normative-factual systems.

When we theorize about factual belief and assertion using traditional truth-conditional semantics, we separate the question, *what are the contents of speech acts and propositional attitudes?* from the question, *what is it to assert, or to believe*

[6] Ibid., 90. [7] Ibid., 93.

something with a certain content? We do compositional semantics on the abstract objects that are provided by the answer to the first question. Without considering what it is to believe a proposition, we can consider how a complex sentence determines a proposition as a function of the semantic values of the component parts. An answer to the second question tells us what it is to be in a certain global state of belief, which determines which propositions are believed by someone in such a state, and what beliefs are expressed in a statement with a certain propositional content, but we can do the compositional semantics without addressing the second question.

It is essential to this strategy that our theory of *propositions* can be characterized independently of any assumptions about the mental states that those propositions are used to describe, or of any account of the speech acts in which propositions are expressed. Our compositional semantics will assume a domain of propositions with a certain structure, and an account of what propositions are. The structure will be a Boolean algebra that can be represented as subsets of a state space, or a set of possible worlds. Propositions can be thought of as properties that the world as a whole might have. If you think of them this way, then *true* propositions can be thought of as the properties of this kind that the world actually does have.

Gibbard does not explain what possible worlds are in a way that is independent of mental states. He asks us to "imagine a god Hera who is entirely coherent and completely opinionated.... [T]here is a completely determinate way *w* she thinks the world to be."[8] I took this at the time, and I think it should be taken, to be just a rough heuristic way of getting a handle on what a possible world is. Propositions, including maximal propositions, can stand on their own as features that reality might have.

If we can understand propositions and possible worlds as abstract objects that are intelligible independently of assumptions about intentional mental states, then we can also understand *complete systems of norms* as abstract object that are intelligible independently of any agent who accepts a system of norms, and even of any account of what it is to accept a system of norms. A system of norms is "a system of permissions and requirements... what matters about a system of norms is what it requires and permits in various conceivable circumstances."[9] Since what is permissible or required, according to such a system, are actions, feelings, or attitudes of persons, we can take the "conceivable circumstances" to be possible worlds, centered on a particular person and time in the world. Then a system of norms can be represented by a function from circumstances of this kind to a set of possible worlds that is permissible in those circumstances, according to that system of norms. Gibbard is clear and explicit that a system of norms can be understood independently of whether anyone accepts the system. "We can characterize any system *N* of norms by a family of basic predicates, '*N*-forbidden,' '*N*-optional,' and '*N*-required'... These predicates are descriptive rather than normative: whether a thing, say, is *N*-permitted

[8] Ibid., 95. In a footnote, Gibbard attributes this conception of a possible world to me, but the contrast he is making is between my view and David Lewis's modal realism. It is not part of my view that possible worlds should be understood as maximal belief states.

[9] Ibid., 87.

will be a matter of fact. It might be N-permitted without being rational, for the system N might have little to recommend it."[10]

So, we have a set W of possible worlds, and a set S of normative systems, defined in terms of those possible worlds, and so a set of all the pairs consisting of a possible world and a normative system. If factual propositions correspond to sets of possible worlds, then factual-normative propositions will correspond to sets of pairs of this kind, and these sets will be the *contents* of factual and normative judgments. We can do the compositional semantics for normative-factual sentences in exactly the same way that we do compositional semantics for purely factual sentences, using this more fine-grained representation of the possibilities that the sentences of our normative-factual language distinguish between. And we can understand what is said by one making a normative claim, and how one revises one's normative-factual beliefs on accepting a normative claim, in the same way we understand these things for factual discourse. If we model a factual discourse with the help of an evolving set of possibilities representing the common ground of the participants in the discourse, then we can represent a discourse that involves normative as well as factual claims with the help of a more fine-grained space of possibilities representing what is commonly accepted and what is in dispute in that discourse.

In stating the Frege-Geach problem, Gibbard asked, "Can we give a systematic account of how the state of mind a complex normative sentence expresses depends on the states of mind that would be expressed by its components alone?" This suggest that the problem should be solved in the context of a certain semantic program that is well articulated in Mark Schroeder's critique of expressivism:

> We should think of expressivism as committed to an underlying semantic program that looks something very much like assertability semantics. The central ideas of assertability semantics are (1) that the role of semantics is to assign an assertability condition to each sentence of the language, understood as the condition under which it is semantically permissible for a speaker to assert it. (2) These assertability conditions typically or always say that the speaker needs to be in a certain mental state. (3) Descriptive sentences inherit their propositional contents (truth-conditions) from the belief that it is their assertability condition to be in. And (4) the assertability conditions of complex sentences are a function of the assertability conditions of their parts, where this function is given by the meanings of the sentential connectives that are used to form the complex sentences.[11]

Schroeder attributes to expressivism a commitment to this kind of semantic program, and the book is an extended argument that it cannot succeed, but on my interpretation of Gibbard's response to this set of problems, he is rejecting this strategy for doing semantics, and instead generalizing the standard truth-conditional semantics that explains content, and the relation between the content of complex sentences and the contents of their constituent parts, independently of the attitudes that those contents are used to describe. As we will see this is not Gibbard's way of understanding what his formalism accomplishes, but it is nevertheless, I will argue, the best way to develop an expressivist semantics.

[10] Ibid., 87. [11] Schroeder 2008, 31.

3. Expressivism about Truth

In the last part of his recent book, *Meaning and Normativity*, Gibbard sets his overall expressivist theory in a context that explicitly rejects the interpretation of normative logic sketched in section 2.

> Frege begins with content... and then invokes different attitudes one can have toward the same piece of content.... I undertake to explain normative thinking in the opposite direction. I begin with a state of mind and then let talk of the content of thinking emerge.... I follow Horwich in treating all mental content as explained by states of mind. I'll speak broadly of "expressivists" in this essay, but mean only expressivists who develop their approach in the ways I advocate. I'll call those who begin their explanations with items of content and explain states of mind as relations a thinker can have with items of content as "Fregeans" (This regardless of whether a theorist accepts the more specific doctrines of Frege, or of Russell, or of anyone else.)[12]

The interpretation of normative logic that I gave in section 2 above is manifestly Fregean in this sense. I will argue that the reversal of the order of explanation that Gibbard proposes cannot work on its own terms, but my main point will be that a defense of an expressivist theory in the spirit of the traditional non-cognitive accounts of normative judgments does not require the broader conception of expressivism that Gibbard articulates in his later work.

Let's return to the myth about Hera "who is entirely coherent and completely opinionated both normatively and factually. She suffers no factual uncertainty; there is a completely determinate way *w* she takes the world to be." This time let's take this not just as a heuristic for getting a handle on the idea of a maximal proposition, or a possible state of the world, but really as a representation of a state of mind. Of course, it is not a state of mind of any actual person—no one is really completely opinionated. We can understand this only as an abstract object—something like the content of a possible state of mind. What explanatory advantage is provided by thinking of these abstract objects as possible states of mind, rather than as maximal propositions (or better as *centered* maximal propositions) that might be the content of a state of mind?

Gibbard is not interested in modeling only fully opinionated states of mind. His aim is to use these abstract objects to model states of mind that are only partially opinionated. The idea seems to be to represent a partially opinionated belief state with the set of all the fully opinionated states that the subject could be in if she learned more, but did not change her mind about anything. But that won't work, since our states of mind will include beliefs about our own present and future beliefs. It is one of my firmly held beliefs that I will *never* be fully opinionated, and I am sure that Gibbard, and all reasonable people who have thought about it, have analogous beliefs. But this means that my belief state cannot be represented by any set of fully opinionated belief states, since there are *no* fully opinionated states that are consistent with it. If we think of the objects used to model a belief state as possible states of

[12] Gibbard 2012, 273.

the world, or maximal propositions we don't have this problem, since a fully determinate state of the world might be a state according to which all agents, throughout their lives, are in only partially opinionated states of mind.

Why is Gibbard suspicious of propositions, thought of as sets of possible worlds, and what advantage does he think one gains if one talks instead about states of belief? I conjecture that the source of his suspicion is that the hypothesis that such objects are contents of belief fails to solve the problem of intentionality—it fails to provide an explanation of what it is about an agent and his or her place in the world that makes it the case that she believes the propositions that she believes. "Some philosophers," he says, maintain "that states of affairs (or 'propositions') can be believed or disbelieved as such, and not just as conceived via one thought or another."[13] I am not sure what it means to conceive of a proposition via one thought or another, or to believe something "as such," but one should not assume that to say that someone believes that ϕ is to deny that there is any story to be told about *how* it is believed: about what the relation is between the believer and the proposition believed. The idea of the strategy that Gibbard calls "Fregean" is to explain content independently of the use to which content is put to describe states of mind. What is assumed by saying that a proposition, in this sense, is the content of an intentional state of mind is only that such states of mind have truth-conditions—they are true under certain conditions, and false under others. A proposition (in the coarse-grained sense defined) is a truth-condition.

At one point, Gibbard says that a version of "the line of thought I am inveighing against treats 'propositions' as sets of possible worlds, *worlds as viewed from nowhere*."[14] But I think Thomas Nagel's oxymoronic metaphor of a view from nowhere is misleading in suggesting that there is something mysterious in the idea of conceiving of a thing as it is in itself. Possible worlds, like anything else, are viewed both by their adherents and by their critics, from the perspectives of those adherents and critics. Saul Kripke, for example, when he wrote his introduction to *Naming and Necessity*, was conceiving of possible worlds from the time and place at which he wrote that introduction, which was somewhere, and not nowhere. But that time and place (or any other distinctive time and place) are not part of what he was conceiving. To use an analogy, it would not be right to describe the moon as being viewed from nowhere because it is not part of what it is to be the moon that we view it from the earth.

However we answer the questions about how believers are related to the contents of their beliefs—of what makes it the case that their states of mind have the truth conditions that they have—it does not seem unreasonable to assume that those states can be described as states with propositional content, and to assume that we have some grasp of the abstract objects that propositions are. I have said that propositions (and possible worlds, which are maximal propositions) are something like *properties* that reality as a whole might have, and this presupposes that propositions are the kind of thing that will be true or false, according to whether the world (reality as a whole) has the property, or does not. Gibbard favors a deflationary or minimalist

[13] Gibbard 2012, 31. [14] Ibid. The phrase comes from Nagel 1986.

conception of truth: to assert that a proposition is true is just to assert it, to believe that it is true is just to believe it. (He is agnostic, but skeptical, about whether there is any more demanding notion of truth.) But I think the minimalist account fails to distinguish between at least two distinct conceptions of truth, which I will label "Aristotelian" and "Protagorean." This distinction matters, I will further argue, for the characterization of expressivism. Gibbard likes to cite the Aristotelian truism, "to speak truly is to say of what is that it is, and of what is not that it is not," and he takes this to be a characterization of a deflationary conception, but what is distinctive about the Aristotelian slogan is that it presupposes a distinction between *what is* and *what is said to be*. The Protagorean conception makes no such conceptual distinction. It can distinguish between what one says (or believes) and what another says (or believes) about what the one says (or believes): Jones can say that what Smith said was false, which will be true in the Protagorean sense for Jones. We, from our perspective, can in turn say that Jones spoke falsely. Of course, we are not *saying* that what Jones said is false *for us*, but we are *expressing* that it is false from our perspective.

The distinction is elusive, since even on the Aristotelian conception, in saying that something is true, one is expressing one's belief, but the theory of propositions, and the derivative notion of a normative-descriptive proposition, help to clarify the distinction. There is a clear and coherent notion of relative truth that is definable within the abstract theory of propositions—truth relative to a possible world. If propositions are modeled by subsets of the space of possible worlds, then they can also be modeled by the equivalent characteristic functions—functions from worlds to any pair of objects {1,0} or {True, False}. A proposition P is true at, or relative to, a possible world w if and only if w is a member of the set P, or if and only if the characteristic function that corresponds to P takes w to Truth. Relative truth is all we need for doing compositional semantics, for the most part, but if our abstract objects are something like properties the world might have, we also want a notion of absolute truth, or equivalently, a notion of the *actual* world. The notion of truth, as a monadic property of propositions, and the notion of a possible world that is actual are interdefinable. Since what we call "possible worlds" are maximal propositions, the actual world can be defined as the maximal proposition that is true. Or if you want your analysis to go the other way, Aristotelian truth can be defined as relative truth, relative to the actual world (that is, a proposition, modeled by a set of possible worlds, is true iff it contains the actual world).[15]

Socrates argued (in Plato's *Theatetus*), that the Protagorean conception of truth was incoherent, but the notion of relative truth that I have defined, in the context of a theory of propositions, is unproblematic. Would it be coherent to hold that relative

[15] There is more than one notion of relative truth. The relativist notion I have defined is not really Protagorean, since it takes truth to be relative to a possible world, while the real Protagorean takes truth to be relative to a state of belief. Gibbard's general strategy of beginning with states of mind, rather than states of the world, would yield a notion of relative truth that is closer to the real Protagorean's view.

I should emphasize that I am not saying that the deflationary notion of truth that Gibbard defends is a notion of relative truth, in either of these senses. Rather, I am saying that the deflationary characterization of truth does not discriminate the absolute notion from the relativist notions.

truth, in this sense, was the only kind of truth we can make sense of? It may be coherent, but I don't think this kind of extreme anti-realism is plausible.[16]

According to Gibbard's normative logic, as I have interpreted it, we refine our possibility space by using norm-world pairs, rather than just possible worlds, as the points in the space on which we define our compositional semantics, and it is straightforward to extend our notion of relative truth to this more refined space. A normative-factual proposition will be a set of norm-world pairs, and such a proposition P will be true, relative to norm-world pair $\langle w,N \rangle$ if and only if $\langle w,N \rangle \in P$. As Gibbard says, "If I oppose stealing ... that commits me to thinking that it is true that stealing is bad. The sense of 'true' here is deflationary, so that 'It is true that stealing is bad' says no more than that stealing is bad." This is fine, so long as the notion of truth we are using is the relativist one. Gibbard adds, "No one, in my view, has explained satisfactorily a more demanding sense of 'true'."[17] I have tried to do this by suggesting that anyone who is a realist about the world of fact should acknowledge a distinction between monadic, or absolute truth, and a notion of truth relative to a possible state of the world. And if we acknowledge this distinction, then we can ask whether the monadic notion applies to factual-normative propositions. Our theory of factual propositions assumes there is an actual world—one maximal factual proposition that is true. But is there also an *actual* system of norms? If we separate, as I have tried to do, conceptions of content (factual, or normative-factual) from any account of what it is to be in a state of mind with content, then we can ask whether the monadic notion of truth, applied to systems of norms, should be part of our theory of factual-normative content. The expressivist should welcome this conceptual resource since it allows us to draw a clear line between expressivists about norms and normative realists. The normative realist answers "'yes" to the question (is there an *actual* set of norms?). On the non-natural realist view, normative claims describe a part of reality as it is in itself. The expressivist should answer "no," since for the expressivist the application of a system of norms is grounded in the states of mind that systems of norms are used to express.

In the last chapter of *Meaning and Normativity*, Gibbard argues that non-naturalism and expressivism may in the end coincide in their theses, but I think his account is unable to distinguish them only because his deflationary notion of truth blurs a conceptual line that a different explanatory framework can use to distinguish the two metaphysical accounts. What Gibbard initially found mysterious in non-naturalism is the idea that "there is a normative realm distinct from the natural realm, and that we have ways to discern how things stand in that realm."[18] The idea was that (according to the non-naturalist realist) a distinction between the natural and the non-natural is a distinction between two more or less distinct and independent domains of reality, a distinction like the one that was once made

[16] David Lewis's modal realism, ironically, might be understood as a version of the view that truth, relative to a possible world, is the only kind of truth there is—at least the only kind of contingent truth. Contingent truth, on Lewis's metaphysical picture, is essentially perspectival. The world that is actual (for us) is just the one we happen to be in. Beings that we are inclined to call "counterfactual" are just beings that are somewhere else.

[17] Gibbard 2012, 20. [18] Gibbard 2012, 235.

between the celestial and the terrestrial, or that one might make between the natural and the supernatural. This view seems mysterious, Gibbard says, because "even if there were such a normative realm detached from all things natural, why would our methods of thinking enable us to discern how things stand in this realm?"[19]

For the expressivist, the distinction between the natural and the normative is nothing like this. "In my picture," Gibbard once said, "all strict facts will be naturalistic.... Apparently normative facts will come out, strictly, as no real facts at all. Instead there will be facts about what we are doing when we make normative judgments."[20] The notion of the "strictly factual" is not a restriction to a special domain of fact, but is used to characterize whatever there is a fact of the matter about, and I take the Aristotelian notion of truth (as contrasted with the useful notion of relative truth) to be an intelligible notion that marks this distinction.

All of the points of agreement between expressivists and non-naturalists that Gibbard notes in his ecumenical chapter are compatible with the interpretation I have given of his normative logic, and with the sharp line marked by the contrasting answer to the question whether there is there a system of norms that is actual in the way that one of the factual states of the world is actual. Gibbard takes Tim Scanlon as his example of a non-naturalist, and quotes him as saying that he finds himself "strongly drawn to a cognitivist understanding of moral and practical judgments.... They obey the principles of standard propositional and quantificational logic, and satisfy (at least most of) the other 'platitudes' about truth." After quoting these comments, Gibbard remarks, "my expressivism lets me agree with Scanlon," and his account of normative logic does explain how he can agree with Scanlon's observations.[21] But Gibbard's minimalism about truth is not a necessary part of the explanation.[22]

It may seem that it is Gibbard's minimalism about truth that allows him to say, "We can speak of the 'world' and normative 'facts' in deflationary senses, so that, in the extreme case, if peas are yucky, then it's a 'fact' that peas are yucky, and this fact characterizes the 'world'."[23] But my claim was that the deflationary notion equivocates between the relativist and the Aristotelian notion of truth. You can, if you like, use the words "fact" and "world" (so long as you retain Gibbard's scare quotes) for notions definable in terms of the intelligible relativist notion of truth, but if you also have the Aristotelian notion, you can make sense of a use of the word "fact" that is restricted to what there is a fact of the matter about.

Most importantly, an expressivist can make the distinction between relativist and monadic truth, and deny (as the emotivists did) that normative judgments are true or false, in the stronger sense, while still retaining the sense in which the content of normative judgments is objective. "Expressivism allows us to interpret the normative conviction that some oughts hold independently of our beliefs and motivations—that one ought not to kick dogs for fun, for example. That one ought not to kick dogs for fun holds true in a possible situation regardless of what anyone thinks or how anyone

[19] Ibid., 236. [20] Gibbard 1990, 23. [21] Gibbard 2012, 230.

[22] Whether or not Gibbard is right that he and Scanlon do not really disagree in doctrine is a further question.

[23] Gibbard 2012, 232.

feels about it, so long as kicking a dog hurts it."[24] What this observation expresses is the acceptance of a system of norms according to which it is not permissible to kick dogs for fun, even in possible circumstances where no one accepts a system of norms that has this consequence. One could accept a system of norms of this kind while also saying that it is not a *fact* that this system is correct.

Gibbard's normative logic is just one example of a general strategy of refining a space of possibilities. A more fine-grained notion of a possibility can be defined by adding a second parameter to the possible-state-of-the-world parameter, which might, as in Gibbard's original exposition of normative logic be a system of norms, but might be some other parameter that relative truth may be defined as relative to. It is useful to have a more general notion of *proposition* that has the same structure as the basic notion of a set of ways the world might be or have been, a notion that may make distinctions between possibilities that the facts do not, in the end, settle. A refinement of this kind lets us generalize the methods that apply to the explanation of factual discourse in two different ways: first, we can extend the methods of compositional semantics to the explanation of the contents of speech acts and propositional attitudes beyond those that aim to represent the way the world in fact is. The generalization is straightforward because the compositional explanation of how the proposition expressed by a complex expression is a function of the propositions expressed by its constituents (with propositions represented as sets of possibilities) can be applied however one individuates the possibilities. Second, we can generalize our pragmatic account of the dynamics of discourse, and of the changing relationships between the attitudes of different rational agents. A discourse (on the kind of model of discourse that I have promoted) takes place against an evolving body of information that represents what is presumed by the participants to be a common background of shared assumptions. These assumptions can include both shared factual information (or misinformation) and shared values or norms. The generalized notion of content helps to show how the structure of a discussion, debate, or cooperative exchange of "information" can take the same form, at a certain level of abstraction, whether what is under discussion is a factual matter, or a question about what to do, or how to think about something. Often, there will be a meta-question in dispute—a question about whether or not there is a fact of the matter about some question. The common structure helps to represent discussions and debates, even at a point at which such meta-questions remain unresolved.

As I have noted, Gibbard argues for a reversal of the order of explanation from the Fregean approach: we should begin with states of mind, rather than with a notion of propositional content. He agrees with Mark Schroeder that the kind of expressivist program that he is advocating takes for granted a notion of *disagreement* between states of mind, and he emphasizes that this is a notion that applies more widely than to the attitude of factual belief. I agree with Gibbard that the task of explaining the broader notions of agreement and disagreement is a central *motivation* for the generalization of the notion of a proposition, but my argument is, the best way to clarify this notion is to distinguish a more general notion of content, from its

[24] Ibid., 233.

application to an account of attitudes and discourse. The distinction seems to me to be intelligible, and it helps to sharpen the Aristotelian notion of truth that is essential to a clear distinction between expressivism and non-naturalist realism.

4. Expressivism about Epistemic Norms

I have focused thus far on very general issues about the distinction between expressivism and non-natural realism, and on the representation of the interaction of factual and normative content. In this final section I will look briefly at some more specific issues concerning epistemic norms with the aims, first of illustrating the way the general normative logic framework works, and second of showing how the framework might clarify some traditional problems about epistemology, and epistemic discourse.

The central general point was that one should characterize normative-factual *content* independently of an account of the attitudes. First, factual content, represented by a set of possible states of the world, is generalized by adding a second parameter, yielding a finer-grained notion of a possibility. Second, one explains the role of the finer-grained content in an account of attitudes, or of discourse. I will first illustrate a version of this strategy in which the second parameter is not, as in Gibbard's theory, a system of norms, but an information state, following some ideas developed in Eric Swanson's constraint semantics, and Seth Yalcin's Bayesian expressivism for epistemic modals.[25] Second, I will look at some issues that arise when we apply Gibbard's specific definition of a system of norms to norms for a Bayesian representation of an epistemic state.

In the Swanson and Yalcin accounts, the extra parameter is an information state, represented by a space of possible worlds and a probability function on it. So, the more fine-grained propositions are sets of pairs consisting of a factual world w and an information state of this kind, S. This refinement allows for a straightforward compositional semantics for epistemic modal expressions such as "might," "must," and "probably." "It might be that ϕ" will be true, relative to $\langle S,w\rangle$, iff the proposition expressed by ϕ (in context) is true in some possible worlds in the information state S. "Probably ϕ" is true relative to $\langle S,w\rangle$ iff the probability of the proposition expressed by ϕ is greater than one half in information state S. But now the second and harder question is, what is the pragmatic role of these "propositions"? What is one doing when one "asserts" one? The idea is that a possibility of this kind represents an information state that it is permissible to be in. A set of possibilities of this kind can represent the common ground—what is mutually agreed—about what is true, and also about what credences it is permissible to have, given the information that we have. In the original, purely factual account, an assertion with propositional content **P** is a proposal to accept **P**, which means to add this information to the common ground. With the more fine-grained propositions, the idea is generalized: an assertion of a more fine-grained proposition of this kind is a proposal to accept (in addition to the factual information entailed by the proposition) a constraint on the

[25] See, for example, Yalcin 2012 and Swanson 2016. See also Moss 2018.

credences that are permissible (according to the information that we share). This simple example illustrates the way the notion of finer-grained content, allows for both a smooth generalization of the compositional semantics, and a smooth generalization of a pragmatic account of discourse. The account counts as expressivist because it takes the assertions of epistemic modal claims as *expressing* epistemic attitudes (that we, the participants in the conversation) are or are not in a position to have certain beliefs or credences), but not as statements *about* our epistemic attitudes. And the account does not require that there be a fact of the matter about what we are in a position to believe, and what credences we are in a position to have.

Now let me suggest how we might apply the Gibbardian characterization of a system of norms to norms for Bayesian epistemology. A system of norms, recall, can be represented by a function taking specific situations that an agent might be in (which may be represented by a possible world, centered on an agent and a time in that world) to a "sphere of permissibility": a set of (centered) possible worlds that are permissible (according to the system) for that agent in that situation. In this case, a world will be permissible iff the agent has a credence function that is permissible in the situation in question. But what are the relevant features of the situation that determines what credences it is permissible to have?

A general expressivist epistemology should have something to say about the norms governing claims to *knowledge*, but I am going to focus here on just one part of the problem by taking *what we know* as part of what defines the conditions relative to which credences are judged to be permissible, or not. Of the propositions not known, but compatible with your knowledge, some are judged much more probable than others. Our question will be about how a system of epistemic norms should assess what *partial* beliefs it is permissible to have in a situation that is in part defined by what one knows in that situation.

It is a general presupposition of the Gibbardian framework that what is permissible is a function of the objective facts about a particular situation, but in order for a rule saying what is permissible or required to be a reasonable rule, the facts that define the conditions for its application must be, normally at least, accessible to the agent. "Add salt when the water boils," for example, is a reasonable part of a recipe only if one can normally tell when the water boils. Still, the recipe does not say "add salt when you *think* the water is boiling." You have failed to follow the recipe if the water was not in fact boiling when you added the salt, even if you thought (and even if you had good reason to think) that it was.[26] The point will apply to a norm that tells you what credences are permitted or required, conditional on what you know. If you are mistaken about what you know in a given situation, you may have credences that fail to conform to norms for credences that you accept, even if you have done your best to conform to those norms. But as with recipes, norms for credence will be reasonable only if agents are *normally* in a position to act on those norms.

So far we have assumed just the following: a system of Bayesian norms can be represented by a function from a specific situation for an agent (one salient feature of

[26] This is a point emphasized by Timothy Williamson, with this example, in his discussion of rules and norms. See Williamson 2000, 223.

which is what the agent knows in that situation) to a set of credence functions that it is permissible for the agent to have in that situation. What more can be said about what a reasonable system of Bayesian norms should look like? We can assume that permissible credence functions will all be ones that assign probability one to all propositions that are known. Among systems meeting these conditions, we can distinguish more permissive from more restrictive norms. At the permissive extreme, a system of norms might allow any coherent credence function that assigns probability one to all the propositions known. At the restrictive extreme, a norm might imply that there is a unique credence function that is permissible in any given situation. It is important not to confuse expressivism in general with permissivism.[27] Epistemic expressivism (as contrasted with normative realism) is the thesis that there is no fact of the matter about what system of epistemic norms is correct. Epistemic permissivism is a thesis about the kind of system of norms we should accept. The two theses are independent: one might be a permissivist and a normative realist, holding that it is a matter of fact that alternative credence functions are permissible in a given situation. On the other hand, one might accept a highly restrictive system of epistemic norms that determines a unique credence function for each situation, while holding, in one's meta-theory, that it is not a matter of fact that any particular system of norms is correct.

Let us assume, for the moment that our epistemic norms are demanding, yielding a unique credence function for each situation. So, we are assuming that the system of Bayesian norms we accept determines a function taking any factual proposition (the agent's total evidence—what he knows) to a credence function defined on a space of possibilities that is compatible with that evidence. One natural version of a function of this kind could be represented by a prior probability function, with the unique credence functions that are admissible in a specific situation defined by conditionalization on the total evidence available to the agent in that situation. It is important to note that the assumption that a system of norms should take this form, while it is a very abstract assumption, is a further constraint on a system of Bayesian norms, one that restricts the relationships between the credence functions that are mandated for different situations. The result of accepting this constraint would be an account of epistemic probability of the kind that Timothy Williamson has given, transposed into a Gibbardian expressivist setting.[28] I want to suggest that the expressivist setting may help to explain the status of an epistemic prior probability function of the kind that Williamson posits.

According to Williamson, the epistemic prior is a measure of "intrinsic plausibility," but he has little to say about where it comes from. It is sometimes said that the Bayesian story presupposes an ur-probability function, representing what credences it would be reasonable to have if one had no empirical evidence at all. According to Alan Hajek, David Lewis used the term "Bayesian Superbabies" for logically perfect rational agents "as they begin their Bayesian odysseys." The credence functions of these hypothetical creatures encode what one's credences should be before receiving

[27] See White 2005 for a characterization and critique of permissivism.

[28] See Williamson 2000, ch. 10.

any empirical evidence.[29] But is it reasonable to assume that the epistemic norms we in fact accept determine what we should believe in a wholly unrealistic situation in which we have had no experience, and have acquired no knowledge at all about contingent matters of fact? I think there is a better way to think about what such a prior might represent, one that derives the norms that an agent actually accepts from the facts about that agent's dispositions to form and endorse credences, and to assess the credences of others.

As Gibbard emphasized in his general account, it is a (natural) fact that an agent accepts a certain (perhaps partial) system of norms. To accept such a system is, in part, to be disposed to do, and to endorse, only what that system of norms permits. It is also a (natural) fact about an agent at a particular time that he has certain credences, and having those credences will be part of what determines the epistemic norms that he accepts. It won't follow that the agent's actual credence function (at a certain time) is thereby compatible with the norms that he accepts at that time for the reason mentioned above: he may be mistaken about what his evidence is. But we should assume that, at least normally, an agent's actual credences, in a given situation will be credences that he is permitted to have, according to norms he accepts, in a situation where what he knows is what he takes himself to know.

The credences of agents change over time as they receive new evidence, including evidence about the credences of others. The way that agents are disposed to respond to possible new evidence, and the way their changing credences are disposed to be influenced by the information received about the credences of others are further facts that it will be relevant to determining what system of epistemic norms the agent accepts. I may ask myself, not just what my credence in some proposition is or should be right now, but what it would or should be under counterfactual circumstances in which I had different evidence than I in fact have. And I may ask what someone else with evidence different from mine should believe. Some questions of this kind will have clear answers for me, others not, but the cases for which the questions have answers (perhaps implicit in my epistemic dispositions, if not in my ability to articulate an answer) will be enough to attribute to me some norms of the appropriate kind.

Some of my beliefs and policies for revising my beliefs are more stable, and more widely shared with others who I take to be epistemic peers, but who may have different specific evidence than I have. Others of my current beliefs and policies are more local, idiosyncratic, and easily revised. (Think of Quine's metaphor of a web of belief, with a stable core, and changes at the periphery.) It may be reasonable to think of a prior probability function as representing the more stable features of my credence function, and such a representation may provide answers to the question of what Bayesian norms I accept.

The kind of prior that we might extract in this way from the credences and belief revision policies of an agent might (and in any remotely realistic case will) assign probability one to some contingent propositions. Suppose contingent proposition **P** has prior probability one. What do my norms say about what credences a person

[29] Hajek 2010.

should have in a situation in which he or she learns that **P** is false? There are two kinds of cases: First, some remote possibilities, while conceivable, are not taken seriously enough for me to have any policies for dealing with possible evidence that they may be actual. That is just to say that our system of epistemic norms may be partial. But second, it may be that sometimes my epistemic norms give me direction even in situations where propositions with prior probability one are discovered to be false. In this kind of case, one receives evidence that one's prior needs to be revised. One may have policies for doing this, and these policies need not be based on some more neutral background prior. One can be a Bayesian without assuming that *every* application of one's norms in a particular situation is based on conditionalizing on a universal prior. It might be (according to a system of norms that I accept) that the appropriate prior for me to have in a given situation is dependent on the facts about that situation.[30]

[30] Thanks to Seth Yalcin for discussion and correspondence, both about expressivism in general and about its application to epistemic norms.

8

Contextualism and the Logic of Knowledge

1. Introduction

Epistemic Contextualism (EC) "is a semantic view, namely the view that 'knowledge'-ascriptions can change their contents with conversational context. To be more precise, EC is the view that the predicate 'know' has an unstable Kaplan character, that is, a character that does not map all contexts on the same content."[1] I will be defending epistemic contextualism, arguing that the case for this kind of account is based not just on judgments about linguistic usage, but on a theoretical account of what knowledge is. But my main aim will be to clarify the thesis: to say how a contextualist analysis of knowledge should be represented, and how we should distinguish the features of the situation in which a knowledge attribution is made that determine what the attribution says from the features that determine whether what it says is true.

I will start, in section 2, with a general discussion of the form that any contextualist analysis should take. In section 3, I will look in detail at David Lewis's contextualist analysis since it is spelled out with his typical clarity and explicitness. While I will argue that Lewis's version of epistemic contextualism is mistaken on several dimensions, it helps to clarify some general issues about the role of context in understanding knowledge claims, and to bring out, by comparison and contrast, some features of the kind of contextualist analysis I want to defend. In section 4, I will sketch a different kind of epistemic contextualism—a version of the information-theoretic account of knowledge—and then in section 5, I will compare and contrast the two kinds of analysis. I will conclude, in section 6, by looking at a puzzle about the dynamics of knowledge and belief, and at the way the kind of contextualist theory I sketch might help to resolve it.

2. The Form of a Contextualist Analysis

I will follow Blome-Tillmann, as quoted above, in characterizing a contextualist analysis of some concept in the framework developed in David Kaplan's *Demonstratives*.[2] For Kaplan, the meaning of a sentence, in general, is a function taking a *context* to

[1] Blome-Tillmann 2009, 244. [2] Kaplan 1989.

a *content*, where a content is a proposition (*what is said* by the sentence in the context). In Kaplan's jargon, the function is called the "character" of the sentence. Singular terms and predicates will also have characters, determining, as a function of context, entities that play the appropriate compositional role in determining the content of the sentences in which they occur. The content will be, or will at least determine, a truth condition. We can model a proposition, as Lewis does, with a set of possible worlds—the worlds in which the proposition is true. A character might be a constant function, taking all contexts to the same proposition (or the same individual or property or relation). The concept in question is context-dependent if the character is not constant: if *what is said* may vary with variations in the context. Context-dependent concepts have what Blome-Tillmann, following Kaplan, calls an "unstable character."

What is a context? Kaplan began by modeling a context simply as an ordered list of parameters, a specification of all the factors that might be relevant to determining what is said in the language being interpreted by the semantics (for example, a time and person representing the time of utterance, and the person speaking, features of context relevant to interpreting tenses and first-person pronouns). He later suggested that a context might better be identified with the concrete situation in which the utterance being interpreted takes place, formally represented by a centered possible world (centered on the speaker and the time of utterance). This would determine all of the relevant parameters.[3] A centered world will determine all of the relevant parameters, not only the objective features of the speaker's environment (such as time and place of utterance) that are relevant to interpreting tenses and pronouns, but also the beliefs, intentions, and presuppositions of the participants in the conversation. Since parameters relevant to interpreting what a speaker is saying in a given situation are normally presumed to be available to the addressees, one might (as I have argued one should[4]) identify the conversational context with the *common ground*—the body of information compatible with the presumed common knowledge of the parties to the conversation.

What is crucial to Kaplan's notion of a context-dependent concept is that the truth of an ascription of the concept depends on the facts in two different ways that need to be distinguished. First, the facts about the context contribute to determining what is said in the ascription. Second, the facts determine whether what is said is true or false. Lewis's contextualist analysis of knowledge does not make this distinction, and does not take the form that Kaplan's framework requires. Instead, the analysis simply specifies a truth-value for a knowledge ascription, relative to a context, where a context includes all the facts that go into determining whether the ascription is true or false. This has caused some confusion in discussions of Lewis's contextualist analysis, but his account can be formulated in a way that does fit the standard format, and doing so will help to clarify the interplay between information that has the role of determining what is said in a claim about a subject's knowledge and information that is the content of such a claim.

[3] see Lewis 1980a. [4] See Stalnaker 2015.

3. Lewis's Analysis of Knowledge

Lewis begins, cagily, with an extremely simple definition that leaves the context-dependence implicit: "Subject *S* knows proposition *P* iff *P* holds in every possibility left uneliminated by *S*'s evidence."[5] "The definition is short," he goes on to say, but "the commentary on it is longer."[6] The commentary requires an explanation of exactly what *evidence* is, and what it is for evidence to eliminate a possibility, which we will consider below, but part of the commentary addresses the meaning of "every," and it is in this little word that Lewis locates his contextualism. He first observes that "every," in general, is a restricted quantifier in most of its normal uses. That is, it ranges over a restricted domain, with the restriction supplied by context. He then says that we need to interpret the "every" in his simple definition in this way.

> If I say that every uneliminated possibility is one in which *P*, or words to that effect, I am doubtless ignoring some of the uneliminated alternative possibilities there are. They are outside the domain, they are irrelevant to the truth of what was said.... Our definition of knowledge requires a *sotto voce* proviso: *S knows that P* iff *S*'s evidence eliminates every possibility in which not-*P*—Psst!—except for those possibilities we are properly ignoring.[7]

Most of the rest of the commentary consists of a defense of various constraints on the possibilities that may be properly ignored.

Possibilities outside of the intended domain—those that are being properly ignored—are, Lewis says, "not relevant to the truth of what was said." If we follow the Kaplanian pattern of analysis, this implies that the restriction on the domain determines what is said—a proposition—which we can then assess independently of the contextual facts that determined that this proposition was the one that was said. So, we might take the analysis to have something like this form:

> Context c determines a class of possibilities **E** that are not properly ignored in c. "*S* knows that ϕ" expresses, in c, the proposition that is true in world *x* iff *S*'s evidence in world *x* eliminates every not-ϕ possibility in **E**.

But this can't be right for the following reason: suppose I say, "Jones knows that there is a book on the table," properly ignoring the possibility that a clever illusion with mirrors made it appear to him that there was a book on the table. The situation was an ordinary one, so I was right to ignore this far-out possibility, but now consider a counterfactual situation in which this properly ignored possibility was realized. Would what I in fact said when I said, "Jones knows that there is a book on the table" still have been true in this counterfactual scenario? The suggested reconstruction of Lewis's analysis implies that the answer is yes, but this answer must be wrong, since knowledge is factive. It is not just that every proper context in which someone says 'S knows that ϕ' entails that ϕ. It is that what is said in every proper context—the proposition expressed—entails that ϕ.

As has been noted in the literature,[8] an analysis like the one I have suggested is a misinterpretation of Lewis for a reason that we can bring out by looking more closely

[5] My discussion of Lewis's analysis draws on Salow 2016.
[6] Lewis 1996, 551.
[7] Ibid., 553–4.
[8] See Holliday 2015 and Salow 2016.

at the way context restricts quantifiers, in general. It is not that context determines the *extension* of the restricted domain; rather, it determines a property that determines the extension of the domain, relative to the world of evaluation. So (to use one of Lewis's examples of a quantifier domain restriction), suppose I say, "Every glass is empty, so it is time for another round." Lewis remarks, "Doubtless I and my audience are ignoring most of all the glasses there are in the whole wide world throughout all of time." In this example and others, what is relevant to determining the proposition expressed with the restricted quantifier is not just the set of glasses that are relevant in the actual world, but also what the extension of the domain would be in counterfactual situations in which we might evaluate the proposition expressed by "all the glasses are empty." The relevant domain, it seems reasonable to assume in this example, is the set of glasses of the people in the group at the table. Suppose Jones usually comes, but isn't there tonight. I say, "If Jones were here, it's likely every glass wouldn't be empty, since he always nurses his one beer all evening." There is no context shift here: the same contextual restriction applies in the scope of the counterfactual supposition. Jones's glass is not in the extension of the intended domain, but it would be if he were here. Or suppose we have invited a group of people to give talks at a conference, and I say, "Everyone has accepted." I mean everyone who was invited, and even if you do not know who was invited (and so don't know the extension of the intended domain of the quantifier), you understand perfectly what I said. It implies that if Smith was invited, then she accepted. All you need to know to understand what I said is that by "everyone" I meant everyone who was invited. So what context determines is not an extensional domain, but a function taking the world of evaluation of the proposition expressed to a domain. The same should be true for Lewis's implicitly restricted quantifier in his analysis of knowledge. What context determines is a property of being a properly ignored possibility, where what exemplifies that property will vary with the world in which we are evaluating the proposition expressed in the knowledge claim. The extension of this context-dependent property will depend in part on which possibilities are being attended to in the context of attribution, but also in part on properties of the subject of the knowledge claim in the world in which we are evaluating what is said in the knowledge claim.[9]

So, what our reconstruction of Lewis's analysis should have been is something like this:

Context c determines a property **P** of possibilities—the property we label "not being properly ignored." "*S* knows that ϕ" expresses, in c, the proposition that is true in world x iff *S*'s evidence in world x eliminates every not-ϕ possibility that has property **P** in x.

[9] It must be granted that Lewis's terminology, "worlds that are properly ignored" is highly misleading on this interpretation, since it suggests that the attributors are doing something improper if they ignore a possibility that is not properly ignored, which in turn suggests that it is a fact about the context of attribution that any given possibility is or is not properly ignored. But the interpretation fits with a general feature of restricted quantifiers, and it is required to make sense of Lewis's proposal, given his characterization of what it is for a possibility to be properly ignored, which explicitly is said to vary with the world of evaluation of the knowledge claim.

On this revised reconstruction, the problem raised for our first suggestion does not arise.

Formally, the property **P** of possibilities in our reconstruction of Lewis's contextualist analysis can be represented by a function taking a possible world *x* to a set of possible worlds—those not properly ignored, relative to world *x*. Equivalently, we can represent it with a binary relation. Let's use 'I^c' to express this relation, so that for context c and worlds *x* and *y*, 'xI^cy' says that in context c, *y* is not properly ignored relative to *x*.

The main tasks of Lewis's commentary on his analysis were, first, to specify how the property of being properly ignored is determined by the context of attribution, and second, to explain what it means for a knower's evidence to eliminate a possibility. He does the first (roughly and informally) by listing some constraints, which are mostly specifications of what possibilities cannot be properly ignored, relative to a given world. The most important of these rules are (1) the rule of actuality, (2) the rule of attention, and (3) the rule of resemblance. The rule of actuality is that the actual world can never be properly ignored, and Lewis makes clear that by "the actual world" he means the world of evaluation of the knowledge claim (as contrasted with the world in which the attribution is being made). So, if we are assessing a claim about what S would know under certain counterfactual conditions, "the actual world" means the relevant counterfactual world. The rule of attention is just that worlds not in fact being ignored (in the context of attribution) are thereby not *properly* ignored. This is the one rule that cannot vary with variations in the world of evaluation. The rule of resemblance is that possibilities that are relevantly similar to those that meet the other conditions are also not properly ignored. It is this rule that does most of the work in showing why Gettier cases are not cases of knowledge.

There is some debate about the interpretation of the rule of attention—what it means to ignore a possibility. Lewis thinks that it is enough to ensure that a possibility is a relevant alternative that the participants in the context of attribution take some notice of it. All a skeptic need do is mention a skeptical scenario, and it is thereby not ignored, and so not properly ignored. This, for Lewis, is what explains the power of skeptical arguments, and the elusiveness of knowledge: if an interlocutor gives an argument for skepticism by describing a scenario that the subject cannot rule out, that will inevitably create a context in which the scenario is a relevant alternative, and so a context in which the skeptic's conclusion is true. Others, such as Blome-Tillmann who defends a variant of Lewis's analysis, think the rule of attention should apply only to possibilities that the attributors "take seriously." Skeptical arguments are more resistible on this variation: one might take notice of a skeptical scenario, but reasonably dismiss it.

Lewis's more skepticism-friendly view has this feature: it cannot be a matter of contention whether a possibility is being attended to. If a participant in a conversation mentions an alternative possibility, then it is manifest that it is being attended to, and there can be no reasonable disagreement about that. On Blome-Tillmann's alternative view, it might be disputable whether a possibility that has been put on the table should be taken seriously. But on a contextualist account, what would the nature of such a dispute be?[10] For the contextualist, the role of the possibilities we are attending to is to determine what the speaker is saying in making a knowledge claim, and not to determine whether what is being said is true. Suppose Alice wants to say that Carl's evidence is sufficient to rule out all non-ϕ possibilities from *this* set of alternatives (the ones she proposes to take seriously), and she makes clear that that is what she is saying. Suppose Bert doesn't disagree with *this* claim, but thinks that certain possibilities that Alice is manifestly not taking seriously *should* be taken seriously. Bert might be tempted to put his objection by saying that Carl doesn't *really* know that ϕ unless his evidence also excludes these other possibilities, but that would be to locate the dispute on the other side of the line that the contextualist draws. That is, it would be to see it as a dispute, not about what is being said, but about whether what is being said is true. Perhaps Carl (if he accepts the contextualist analysis) could put his disagreement this way: "I don't disagree with what you are saying, but I think we should be asking a different question: whether Bert's evidence rules out all non-ϕ possibilities from this wider set of possibilities."[11]

While Lewis insists on his strict interpretation of the rule of attention, he acknowledges that in practice we bend the rules. "What if some far-fetched possibility is called to our attention, not by a skeptical philosopher, but by counsel for the defence? We of the jury may wish to ignore it, and wish it had not been mentioned. If we ignored it now we would bend the rules of cooperative conversation; but we may have good reason to do exactly that."[12] It is not clear to me what the difference is, at least in cases where there is no given rulebook, between a strict rule that we are allowed to bend and a rule that allows for flexibility.

As for the second task of Lewis's commentary on his analysis (explaining what evidence is, and how it eliminates possibilities), here is what he says:

> The uneliminated possibilities are those in which the subject's entire perceptual experience and memory are just as they actually are. There is one possibility that actually obtains (for the subject and at the time in question); call it *actuality*. Then a possibility W is *uneliminated* iff the subject's perceptual experience and memory in W exactly match his perceptual experience and memory in actuality.[13]

[10] Aside from the specific issue about the interpretation of Lewis's rule of attention, there is a more general problem about contested contexts. A context, in the kind of account I favor, is modeled by a body of information, which is presumed to be shared by the participants to the conversation. But there may be disagreements between the participants about what they are in a position to accept as common ground, and negotiations about what the common ground should be, and for that we need a context in which those negotiations take place. I discuss this general issue (especially as applied to epistemic modals) in chapters 7 and 8 of Stalnaker 2015. See also DeRose 2004 for a discussion of contested contexts for knowledge attributions.

[11] The dialectic is complicated here. If Bert is giving an argument for skepticism, then his reason for insisting that Carl must be able to rule out certain far-out possibilities is that he thinks that it is not just Carl, but also they themselves—Alice and Bert, the parties to the context of attribution—who are not in a position to rule out those possibilities. If a far-out possibility might (for all the attributors know) be the actual situation, then it cannot be ignored. But there might be a dispute of this kind about whether the *subject* is in a position to rule out a possibility even in cases where the parties to the context are agreed that *they* are in a position to rule it out.

[12] Lewis 1996, 560. [13] Ibid., 553.

One might, Lewis says, be tempted to interpret this notion in the following way: One might take perceptual experience and memory to have a propositional content: the way things appear to the subject to be, and the way the subject seems to remember his experience having been, and one might take this propositional content to be the evidence that eliminates possibilities. But this would be a mistake. "When perceptual experience *E* (or memory) eliminates a possibility *W*, that is not because the propositional content of the experience conflicts with *W*.... Rather it is the existence of the experience that conflicts with *W*: *W* is a possibility in which the subject is not having experience *E*."[14] Lewis is here following Quine, who emphasized that the empirical foundation for knowledge ("surface irritations," in Quinean rhetoric) should be described in physicalistic terms, and distinguished from the protocol sentences of a phenomenalist language. The data, in this sense, do not provide a conceptual foundation for knowledge—they are not the subject's reasons or justification for the beliefs that constitute her knowledge. Nevertheless, evidence, for Lewis, is propositional: propositions are sets of possibilities, and the evidence that eliminates possibilities, on Lewis's account, is the proposition that includes all and only worlds in which the subject is having experiences that match those in actuality. And, for Lewis, evidence is knowledge[15]: The analysis implies that the subject knows the evidence proposition, and knows it in every context in which the contextualist analysis might be applied. Lewis dismisses the idea that "experience has some sort of infallible ineffable purely phenomenal propositional content.... Who needs that?" But so far as I can see, his analysis is committed to *propositions* that are, in a sense, phenomenal (they are about the internal experiential state of the subject at the time) and that are infallible in the sense that the analysis ensures that they are known, no matter what possibilities are or are not properly ignored.

On some contextualist analyses, such as one I will defend below, what counts as evidence will itself vary with context, but it is an essential feature of Lewis's account that evidence is context-invariant, and so in any context knowledge rests on an invariant foundation. It is also a manifest feature of Lewis's characterization of the way evidence eliminates possibilities that the relation is symmetric: the evidence S has in *x* eliminates possibility *y* if and only if the evidence S has in *y* eliminates possibility *x*.[16] The elimination relation is the complement of an equivalence relation:

[14] Ibid.

[15] I am not saying that Lewis subscribes to Timothy Williamson's thesis, E = K (evidence is identical to knowledge), which he manifestly does not, but just the weaker thesis that all evidence is knowledge. Lewis endorses (implicitly) just two of the three theses that are the premises of Williamson's argument for E = K: that evidence is propositional, and that evidence propositions are known. He does not and could not (given his conception of evidence) endorse Williamson's third and more controversial thesis that all propositions that are known by a subject count as part of his evidence. (Williamson's E = K thesis will, however, be true on Lewis's account in the maximal context in which no possibilities are properly ignored.)

[16] In some contrasting ways of understanding evidence, this relation is not symmetric, and this non-symmetry plays a role in critiques of skeptical arguments. For Williamson, for example, the evidence S has in a skeptical scenario does not eliminate the normal scenario—that is why he is deceived in that scenario, thinking that everything is normal. But (according to Williamson) the evidence S has in the normal scenario *does* eliminate the skeptical scenario. The asymmetry allows us to say that S has the knowledge, in the normal scenario, that things are as they appear to be.

say that $x \approx_S y$ if and only if the evidence S has in world x is the same as the evidence S has in world y.

Using these two binary relations, we can put our reconstruction of Lewis's analysis into the format of the standard model theory for epistemic logic in the style of Hintikka, according to which there is a binary epistemic accessibility relation R (for each knower and time) and a semantic rule that says that S knows that ϕ in possible world x if and only if ϕ is true in all worlds y such that xRy. Lewis's context-dependent epistemic accessibility relation can be defined this way:

$$xR^c{}_Sy =_{df} (xI^cy \;\&\; x \approx_s y).$$

The epistemic accessibility relation is reflexive on this analysis, since both of the relations in the conjunctive analysis are reflexive. We cannot assume, however, that the relation is either transitive or symmetric, since the relation I^c may not be transitive or symmetric. It is reasonable to think that there will be counterexamples to transitive because of the rule of resemblance, and the intransitivity of resemblance. Suppose both y and z are being ignored in the context, but y is not being *properly* ignored relative to x because it resembles x in relevant respects, and z is not *properly* ignored relative to y because it resembles y. If z does not resemble x, it still might be that z is properly ignored relative to x. There will be a counterexample to symmetry when (and only when) there are worlds x and y such that y is properly ignored relative to x, but x is not ignored in the context, and so not properly ignored relative to any world. In any such case, yI^cx, but not xI^cy.

The I^c relation will, however, have the following *qualified* symmetry property that will be inherited by the epistemic accessibility relation, $R^c{}_S$: if x is not being ignored in context c, then for any y, if xI^cy, then yI^cx. This structural feature of the analysis gives rise to potential counterexamples.[17]

Suppose Carl is in a skeptical scenario (in world x): it seems to him that there is a book on the table, but in fact it is a clever illusion done with mirrors: There is no book on the table. Alice and Bert (the parties to a discussion about what Carl knows) know that this is an illusion, and that there is no book on the table. They are of course attending to world x, since it (or a world like it) is recognized to be the actual situation of the context, and since it is highly relevant to what Carl knows, which is what they are discussing. Carl of course does not, in world x, know that there is a book on the table, but since he is deceived by the illusion, he takes himself to know that there is a book on the table. Or so it seems. But let y be *any* situation that is compatible with what Carl knows in x—any situation in which $xR^c{}_Sy$. It will follow by the qualified symmetry property that $yR^c{}_Sx$. Since there is no book on the table in world x, it follows that in world y, Carl does not know that there is a book on the table. Since y was an arbitrary world compatible with Carl's knowledge in x, we can conclude that it is true in *every* world accessible to x that Carl does not know that there is a book on the table. It therefore follows that Carl knows, in the actual world x, that he does not know that there is a book on the table: he knows his situation is a skeptical scenario. But this consequence of the analysis (it seems) is obviously false.

[17] This point is brought out in Salow 2016.

My example is a version of one used by Bernhard Salow in his incisive paper on Lewis on iteration principles to show that the analysis has this problematic consequence. Salow tries on a bullet-biting response (which he labels, using a different metaphor, the hardnosed response) to the consequence, and like a resourceful defense attorney for an obviously guilty defendant, he succeeds in showing that there is more to be said for biting this bullet than one might have thought.[18] He can agree that it is a datum that the statement, "Carl knows that he doesn't know that there is a book on the table" is false, but his suggestion is that this might be explained by a context shift: in the scope of "Carl knows," the inner "know" is interpreted relative to the subject Carl's private context, rather than relative to the context of attribution. He observes that, according to some analyses there is precedent for a context shift in the scope of an attitude ascription (for example a shift in the interpretation of epistemic "might," or of predicates of personal taste, such as "fun" or "tasty"), but this is controversial, and it seems a stretch to hypothesize such a shift in the case of the interpretation of "S knows" within the scope of "S knows." And if there is *some* sense in which Carl knows about his ignorance, it ought to be possible to find a way to say it. For example, one might (if Lewis's analysis were correct) expect the following to be true in Alice and Bert's context: "It is a fact that Carl doesn't know that there is a book on the table, and he knows that this is a fact." But, intuitively, there does not seem to be any sense in which he has this knowledge. Alice might say that Carl *would* know of his ignorance if he took seriously the possibility that he was in a skeptical scenario, but this is explicitly counterfactual. And since Alice and Bert are taking this possibility seriously only because it is the actual scenario (not because their high standards for knowledge require them to take it seriously anyway), Carl (who, we may assume, has the same standards for knowledge as Alice and Bert) would have reason to take it seriously only if he knew that this scenario was actual, and he does not know *that* fact relative to either Alice and Bert's context or his own. So, I think we should take this example at face value, as a counterexample to Lewis's analysis.

The qualified symmetry principle for the epistemic accessibility relation is a feature of the structure of the analysis, and does not depend on the details of the rules that determine what possibilities are properly ignored. It *does* depend on the rule of attention—on the fact that if the attributors are attending to a certain possibility, then it is not properly ignored, relative to any possibility. But this feature of the rule of attention is independent of whether one interprets that rule in Lewis's way (so that any possibility mentioned is thereby not ignored), or in the more restricted way suggested by Blome-Tillmann (according to which the rule applies only to possibilities that are taken seriously by the attributors as live options). So, if we take Salow's example as a counterexample, as I have argued we should, it is a counterexample that

[18] I say he "tries on" this line, since he presents it as an option, but does not in the end endorse it. He suggests that it could be plausible only in cases where the two contexts (Bert and Alice's context of attribution, and Carl's context of self-attribution) differ in their standards for knowledge (high stakes contexts vs lower stakes contexts), and not just in their knowledge of what is actual. But we can stipulate that their standards are essentially the same: for both, possibilities regarded as far-out are not taken seriously. But what is regarded as far-out will inevitably depend on what one knows.

cannot be avoided by varying the details of Lewis's analysis while retaining its structural features. But this does not mean that this kind of example poses a general problem for contextualism about knowledge, or for the general idea that knowledge is essentially contrastive: the capacity to distinguish the actual world from certain relevant alternatives, determined by context. The example does make salient the general problem for a contextualist analysis of distinguishing the interacting features of the situation of the attributor of knowledge and the situation of the subject to whom knowledge is being attributed. In the remainder of this chapter, I will consider a different way of developing a contextualist account that I hope will help to clarify this interaction.

4. The Information-Theoretic Model of Knowledge

Knowledge, it seems reasonable to assume, is a capacity to make one's actions depend on certain facts. This requires that an agent be capable of being in states that *carry information*, and that are available to be used to determine one's actions. On the information-theoretic picture, this means that there are states of the agent that vary systematically (under certain normal conditions) with corresponding states of the environment. I will start with a simple model (One might say, as Paul Grice did about one of his highly complex proposed analyses, that it is a "hideous oversimplification"), but I hope it will bring out some features that a more realistic model will share. Assume that the agent may be in states $I_1, \ldots I_n$ and the environment may be in states $E_1, \ldots E_n$ that correspond in the following way: there are conditions N such that for all j, $N \vdash I_j \leftrightarrow E_j$. In a simple model of this kind, if conditions N obtain, then agent S knows that ϕ iff for some j, (1) S is in state I_j, and (2) $N\&I_j \vdash [\![\phi]\!]$.[19]

But what happens, on this kind of account of knowledge, when conditions N do not obtain? If we were talking about the kind of simple and inflexible information-carrying devices often used as toy models of the information-theoretic conception of knowledge (such as thermostats or fuel gauges), the answer would be that the attributor describing what is "known" would be presupposing that the normal conditions obtained. Such information-carrying attributions require a context in which they are presupposed to obtain. With such devices, the functional organization relative to which the informational content is defined will be fixed, and the range of information the device is capable of carrying will be limited. When the channel conditions for such a device fail to obtain, the question of what it knows does not arise. The simple thermostat does not "know" whether it is functioning correctly—whether its channel conditions in fact obtain or not, and when they do not obtain, there is nothing it can be said to know, although questions about what it *indicates* may still be asked and answered. The defective thermostat is in a functional state analogous to false belief, a state that would carry certain information if it were functioning correctly. But we are interested in more complex devices with a richer range of potential channels of information, and this will require a more complex structure for determining normal conditions.

[19] We use the double brackets, enclosing a sentence, to represent the proposition expressed.

To have just a slightly more complicated model, we might suppose that information comes through multiple channels, each with their own normal conditions.[20] Conditions are *fully* normal only if all channels are functioning normally, but when some conditions are normal and others not, then the agent will still receive the information that comes in through the channels that are functioning normally. In this more complex model (still a hideous oversimplification, but a move in the direction of realism), there will be a set of propositions, $N = \{N_1, \ldots N_m\}$ stating the normal conditions for the different channels through which information may be received, and the analysis will say that S knows that ϕ if for some subset of N, all the propositions in that subset obtain (so all of those channels are functioning normally), and their conjunction entails that $(I_j \rightarrow \phi)$. On one variation on this more complex model,[21] some of the members of N might form a sequence of propositions of decreasing strength, representing more and less stringent normality conditions of the same kind. The subject's internal state might carry more accurate information under some environmental conditions, and less accurate information under others. For example, an oven thermometer might measure to a margin of error ± 1 degree under stringent conditions, and to a wider margin of error, ± 3 degrees, under less stringent conditions. If only the less stringent conditions are in fact met, and the thermometer registers 300 degrees, then S will not know that the temperature is 300 ± 1, but will still know that it is 300 ± 3.

There might in addition be norms governing the interaction of information received thorough different channels—normal vs abnormal ways in which complex information received from different sources is correlated, and is processed. Some conditions will distinguish normal from abnormal features of the environment while others will concern the internal workings of the subject's perceptual or cognitive systems. Some conditions will concern local channels by which information is received, while others will be more global constraints on the normality of the environment in which information is received, and more general conditions on the normal functioning of the knower's cognitive capacities. Some conditions will include very local prima facie generalizations about the environment as in the following example: Alice has the capacity to distinguish by sight her son Carl from everyone except a person Dan, who lives far away, and who she knows nothing about, but who happens to be an exact double of her son. Her sighting of Carl approaching her house still carries the information that it is Carl that she sees, since a counterfactual scenario in which it is Dan who is approaching is an abnormal possibility, and so not a relevant alternative. Of course, if it really were Dan approaching, this situation, even though an abnormal possibility, would be a relevant alternative, since it would be compatible with those normality conditions that actually obtain. Even if it is Carl who is approaching, if Dan happens to be in the neighborhood—perhaps he walked by just a few minutes before—that might be enough to make the situation abnormal, and so make Alice's sighting of her son a sort of Gettier case. (Dan would be like the proverbial fake barns.)

[20] This kind of model is discussed briefly at the end of Stalnaker 2006, reprinted as chapter 1 of this book.

[21] This variation was suggested in discussion by Dan Hoek.

Local skeptical scenarios—tricks with mirrors, bad lighting conditions, or the appearance of improbable doppelgängers—will be cases where more local normality conditions are violated, but more general conditions about the way we are normally connected with our environments will still be satisfied. So even when it is Dan that Alice sees, she will still know that she sees a boy approaching. Global skeptical scenarios involving brains in vats, evil demons, or coherent hallucinations will be cases where even these more general conditions fail.

However the details are filled out, a complex structure of normal conditions suitable for more complex information carriers and processors such as ourselves can be assumed to determine an ordering structure yielding more and less great deviations from ideal normality. The structure will determine, for each possible world x, a proposition N_x that gives the normal conditions that are satisfied in world x. Then we can say that the subject knows that ϕ in world x iff for some j, he, she, or it is in internal state I_j, and $N_x \& I_j \vdash [\![\phi]\!]$.

The structure of normal conditions is determined by the attributor's presuppositions—it is the attributor who is claiming that the subject can discriminate the actual situation from certain relevant alternatives. But as with Lewis's context-restricted quantifiers, there are constraints, and the epistemic situation of the subject is relevant to determining what the admissible normal conditions are. In particular, while neither the attributor nor the subject needs to have considered the proposition that a certain normality condition does or does not obtain, if a certain condition *is* considered, and is doubted, by the *subject*, then this will be enough to ensure that that condition is not a normality condition for that subject at that time. Consider a standard example: suppose that L, *lighting conditions are normal*, is true in a certain situation, but that Alice (for either good reason, or for no reason) doubts that they are. As a result, she fails to believe that the wall is red, even though it looks red to her, and so she fails to know it. But since the wall looks red to her, it remains true that Alice is in a state that meets the following condition: if lighting conditions were normal (and her perceptual system were functioning normally), she would be in that state only if the wall were red. So, if L were one of the normality conditions relative to which her knowledge is determined, it would follow that she knew that the wall was red. Since she does not, L cannot be one of the normality conditions for that context.

Jonathan Vogel has used the term "subjunctivism" for "the doctrine that what is distinctive about knowledge is essentially modal in character, and thus is captured by certain subjunctive conditionals."[22] The information-theoretic account I have sketched might be called a version of subjunctivism since the ordering structure it appeals to is the kind of structure that has been used to interpret counterfactual conditionals, and since the relations the orderings reflect are relations of causal and counterfactual dependence and independence. But our account contrasts with the paradigms of subjunctivist analyses such as those of Dretske and Robert Nozick in that the appeal to counterfactuals is indirect. Nozick's proposed analysis[23] says what it is to know that ϕ directly in terms of counterfactuals involving the proposition that ϕ. With this kind of analysis, the non-monotonicity of the counterfactual results

[22] Vogel 2007, 73. See also Vogel 1999. [23] Nozick 1981.

in a failure of logical closure conditions for knowledge. But our analysis uses the ordering structure to determine a holistic knowledge state, and then says what it is to know that ϕ in terms of the properties of that state. This analysis, like Lewis's, can be represented as a special case of the standard modal analysis, with a binary epistemic accessibility relation. Propositions known in possible world x are those entailed by the proposition that is the set of possible worlds accessible to world x.

Wesley Holliday, picking up on Vogel's terminology, describes a family of logics of knowledge that he calls "subjunctivist-flavored," that are appropriate for Dretske's and Nozick's accounts, and that deviate from the standard modal formula in a way that results in a failure of deductive closure conditions.[24] We will say more about closure conditions below, but let's look first at the similarities and differences between Lewis's analysis and the normal-conditions analysis, both of which conform to the standard formula, and so satisfy the closure conditions.

5. Comparing the Two Analyses

First, the two analyses are both contextualist: the truth-conditions for attributions of knowledge may vary with variations in the contextual presuppositions of the attributor. On Lewis's analysis, which possibilities are properly ignored, relative to a possible world, depends on which are ignored, which, however that is interpreted, is wholly a matter of the context of attribution. On the normal-conditions analysis, the structure of normal conditions and the partitions of possibilities that are correlated under those conditions will depend on empirical facts about the causal relations that hold, but the distinction between which situations are normal and which are deviations from the norm will be different in different contexts of attribution. Contextual variation is essential, since we can ask about any specific normal condition whether the subject knows that *that* condition obtains, and addressing this question will require a context in which the normality of that condition is not presupposed. And even given a structure of normal conditions, in some contexts it will be presupposed by both attributor and subject that certain of the more stringent normal conditions do not obtain, and so they will not be relevant, either to the attribution of knowledge, or to attribution of full belief.

There is this important difference between Lewis's contextualism and that of the normal-conditions analysis: On Lewis's account, there is a default context—the maximal context in which no possibilities are properly ignored—and the analysis gives a precise account of what is known relative to that context. Furthermore, every knowledge claim made in any Lewisian context c will be equivalent to a knowledge claim with a conditional content, interpreted relative to the default context. Specifically, let c be any context, and let ψ expresses the proposition true in all of the possible worlds that are properly ignored in context c, relative to world x. Then for any ϕ, S knows that ϕ in world x, relative to context c iff S knows that $(\psi \rightarrow \phi)$ in world x relative to the default context. So, all context-dependent knowledge claims can be expressed in a context-independent default context that requires no

[24] Holliday 2015.

contingent presuppositions. In contrast, on the information-theoretic picture, there is no context for the attribution of knowledge that does not require any contingent presuppositions. Any candidates for the channel conditions for the transmission of information will be contingent propositions about the causal relationships between the knower and the facts known, and any question about whether the agent knows that those conditions obtain must be asked in a context where alternative normal conditions are presupposed to hold.

There are some purely formal similarities between these two contextualist analyses that help to highlight their formal differences. In both cases, the epistemic accessibility relation is defined as a conjunction of two binary relations between possible worlds: one defines the relevant alternatives to world x, for each world x, and the other says which of those alternatives are ruled out by the evidence that the subject has in world x. At this level of abstraction, the two analyses are alike. The differences will be in the differences between the two binary relations used to define epistemic accessibility.

Recall that in our reconstruction of Lewis's account, $xR^c{}_S y$ iff $(xI^c y \;\&\; x \approx_S y)$, where I^c is the relation of being properly ignored: As we noted in our discussion of Lewis's account, the I^c relation is reflexive (because of Lewis's rule of actuality), and it satisfies the qualified symmetry condition that gave rise to counterexamples. We also noted that this relation cannot be assumed to be transitive, and that because of Lewis's rule of resemblance it seems reasonable to assume that it will not be transitive.

On the normal-conditions analysis, the relation that corresponds to I^c is defined in terms of a structure that orders the possible worlds. Let's use "N" for this relation: $xN^c y$ iff all of the normal conditions determined by the context that are satisfied in world x are also satisfied in world y. The ordering will be a partial order, so the relation will be transitive and reflexive, but not necessarily connected. The relation will also be strongly convergent,[25] since there will be worlds in which *all* the normal conditions are satisfied, and these will stand in the relation to any world that is N-related to world x.[26] It will not, however, satisfy the qualified symmetry condition, so will not face the kind of counterexample that Lewis's analysis faces.

The second components of the analyses of epistemic accessibility are, in both the Lewis and the normal-conditions accounts, equivalence relations, but the most important contrast between the two analyses is in the way this relation is understood. The relation in Lewis's analysis is context-independent, since the restriction on possibilities that are relevant alternatives plays no role determining the relation of phenomenal indistinguishability. Lewis says, "I say that the uneliminated possibilities are those in which the subject's perceptual experience and memory are just as they

[25] A binary relation R is strongly convergent iff for all x there is a w, such that for all y and z, if xRy and xRz, then yRw and zRw.

[26] One might question whether it is always compatible with the subject's knowledge that all normal conditions be satisfied, since the subject may discover that conditions are, in some respect, abnormal. But when what *was* a presupposed normal condition in a prior epistemic situation is called into question by the subject, it is no longer a relevant normal condition in the posterior context. (Thanks to Bernhard Salow for discussion of this issue.)

actually are," and he makes clear that he means to be identifying a relation that is not intentional. It is the having of the experiences, and not any content that they may have, that is shared in worlds that stand in this relation. The corresponding relation in the normal-conditions analysis, however, is determined by the functional organization that attributors of knowledge and belief are presupposing, and so is defined relative to their context. The causal structure of any application of that functional theory hypothesizes a linked pair of partitions of the possibility space—one a partition of propositions about the environment that the subject's knowledge is about, and the other a partition of possible states of the subject that are available to control his, her, or its actions. The relevant internal states determined by the partition are not phenomenal states, and will be individuated differently for different hypotheses about the overall functional organization, and in particular on the external and internal conditions that are the normality conditions for that context. So unlike Lewis's internal states, they are individuated intentionally—by the content that they determine, under normal conditions. The "internal" states are belief states, and since they are characterized in terms of the external conditions that govern the transmission of information, they are states that are external in the way that Putnam's and Burge's thought experiments showed that meaning and belief are states that depend on their environment.[27] The analysis says that being in state I_k carries the information that the world is in state E_k, provided that the conditions N in fact obtain. But whether they obtain or not, the state I_k will carry the information that *if* N obtains, then the world is in state E_k.

The contrast between Lewis's conception of the internal states and that of the information-theoretic analysis is connected with a very general epistemological contrast. For Lewis, the internal states are the knower's *evidence*: they are the foundation or the basis for the subject's knowledge. His analysis is a version of classical foundationalism according to which there is a part of one's knowledge that is especially secure, and that is the basis for the rest of what one knows. But it would be a serious mistake to think of the information-theorist's internal states as evidence that plays this kind of role. The internal states, individuated by their content, are belief states: I_k is the state of believing that the world in in state E_k. But having this belief is not a part of my knowledge that grounds or justifies the rest of the knowledge that I have in virtue of being in that state. That gets things backward. The information-theoretic analysis fits better with the Williamsonian "knowledge first" picture according to which the concept of belief is derivative from the concept of knowledge, and anything that one knows can be evidence.[28]

But let's go back to the structural similarities between the analyses. Since they both define epistemic accessibility with a conjunction, we can, in each case, distinguish

[27] See Stalnaker 1993, where I used the information-theoretic story of intentionality to give a theoretical defense of Tyler Burge's anti-indvidualist thought experiments. The situations described in those experiments were counterfactual situations in which conditions were abnormal with respect to the actual world, but normal with respect to those situations themselves.

[28] Greco 2017 defends a contextualist version of foundationalism that allows that evidence might be a proper part of one's knowledge that is, in context, the foundation for the rest, but that does not play this role in another context, where it may be doubted or be in need of evidential support. The information-theoretic story may be compatible with such a view, but it will not be the internal states that play this role.

two classes of propositions that are true in all worlds that are epistemically possible, relative to world x. First, there are those that are true in all relevant alternatives (true in all worlds that stand in the first relation, I^c or N^c, to x). Second, there are those that discriminate between relevant alternatives (false in some worlds that stand in the first relation, I^c or N^c, to x, but true in all worlds that stand in both the first and the second relations). Both count as propositions that are "known" according to the classical analyses, since the propositions of both kinds are true in all epistemically possible worlds, but those in the first are "known" simply in virtue of the limits on the set of possibilities that the knower is presupposed to be distinguishing between. Knowing these propositions is no kind of cognitive achievement. To use Holliday's terminology, they are known "vacuously."

Lewis observed that what we have called vacuous knowledge counts as knowledge on his analysis, but suggests that this is unavoidable. "If we analyze knowledge as a modality, as we have done, we cannot escape the conclusion that knowledge is closed under (strict) implication." "I have to grant, in general, that knowledge just by presupposing and ignoring *is* knowledge."[29] But given the simple and clean distinction between the two kinds of propositions that are "known," it is not at all clear why this consequence cannot be avoided by a superficial change within the context of this kind of modal analysis. Say (in Lewis's framework) that S knows* that ϕ iff $\{y: xR^cy\} \subseteq [\![\phi]\!]$, but not $\{y: xI^cy\} \subseteq [\![\phi]\!]$. Or in the Information-theoretic analysis, say that S knows* that ϕ iff $\{y: xR^cy\} \subseteq [\![\phi]\!]$, but not $\{y: xN^cy\} \subseteq [\![\phi]\!]$. To identify knowledge with knowledge* would be a very minor change in either of these overall theories, and it would fit with the idea, common to both contextualist pictures, and emphasized by Dretske, that knowledge is essentially contrastive: a capacity to distinguish the actual world from certain relevant alternatives.

Analyses with a structure like knowledge* are familiar in modal semantics for wishing and wanting, and as a response to some puzzles about deontic operators.[30] I wish I knew who committed the murder, which (on a modal analysis) means that my wishes would be satisfied only in a possible world in which I know who committed the murder. But the murder was committed in all of those worlds, and that is not something I wish or wished for. That the murder was committed is presupposed, and so true in all possible worlds that my wishes distinguish between, and so true in all of those worlds in which my wish is satisfied. But if wishing that ϕ is true only for ϕ that discriminate between a set of relevant worlds, then the analysis won't have the unwanted consequence. An example Dretske used in a discussion of closure conditions brings out that some epistemic notions need an analysis more like knowledge*: Jones sees that there is wine left in the bottle, but does not see that the liquid in the bottle is wine. He may *know* that it is wine, since he has tasted it, but he doesn't *see* that it is, unless he discriminates, by sight, the wine-left-in-the-bottle worlds from a wider set of worlds in which the liquid in the bottle is not wine.[31] One might add that perhaps he doesn't even know that the liquid in the bottle is wine: it is just that those saying what Jones sees presuppose, reasonably and correctly, that it is.

[29] Lewis 1996, 263 and 262.

[30] I discussed this kind of analysis in Stalnaker 1984, ch. 5.

[31] Dretske 2014, 29.

Suppose Alice sees that the wall is red, and so comes to know that it is. She acquires this knowledge only if lighting conditions are normal and her visual system is functioning normally, but suppose these conditions are met. It will then be true that the wall is red in all of Alice's epistemic alternatives, and also true in all epistemic alternatives that that she sees that it is. (If asked, "How do you know the wall is red?" she might reply, "I see that it is," indicating that she knows not only that the wall is red, but that she sees that it is red.) So, it will be true in all epistemic alternatives that lighting conditions are normal. But does she *know* that lighting conditions are normal? Perhaps she hasn't considered that, and has no evidence about the lighting conditions—she just takes for granted that they are, and that her perceptual capacities are functioning normally. Those who are attributing knowledge to her take this for granted as well. If you are inclined to say that a person need not *know* that perceptual conditions are normal in order to know thing by perception, then you are using a notion more like knowledge*.

I don't want to claim that one or the other of "knowledge" and "knowledge*" is the right account of the way the word is commonly used—probably we go back and forth between them. But however one answers the question whether the analysis should identify the intuitive notion of knowledge with truth in all epistemic alternatives or with something like the contrastive notion, knowledge*, it will be important to distinguish the two categories of propositions that are true in all epistemic alternatives (those the subject knows*, and those that were being presupposed) since they may play different roles in the way the knower responds to new evidence, and in the ways that contexts of knowledge attribution shift. In the next and last section I will consider briefly the role of this distinction in the dynamics of knowledge and belief.

6. Belief Revision

Since knowledge is factive, one cannot discover that something one knows is false, but just as a non-skeptic can consider and reason about the possibility that he is in a skeptical scenario, so anyone can entertain the hypothesis that he discovers that something he took himself to know is in fact false, and can consider how he ought to revise his beliefs in response to that discovery. Belief revision theory is an attempt to model the abstract structure of the policies that an agent has or should have for responding to such a discovery. From the perspective of the information-theoretic conception of knowledge, such a discovery would be the recognition that some normality condition that one was taking for granted must be false, and the revision will involve consideration of what went wrong—what condition presumed to be normal was in fact not.

Whenever one recognizes the falsity of some condition presupposed to be a normality condition relative to which one's knowledge is defined, that forces a change in the context, since new alternative possibilities become relevant alternatives. The standard belief revision theory assumes that if the new evidence is compatible with all of one's prior full beliefs, then all prior beliefs or claims to knowledge are retained. Partial beliefs will change, by conditionalization, but there need be no context shift: no change in the presupposed structure of normal conditions, or the presuppositions about which of the normal conditions obtain. That is, the standard belief revision theory, even in its weakest form, assumes the modest conservative principle that full beliefs are retained in the face of new evidence so long as the evidence is compatible with *all* of them. The rationale for this condition is the assumption that a full belief, or a claim to knowledge, constitutes a commitment to continue believing it unless forced to give something up.

But now suppose an agent's new evidence is compatible with everything presumed to be known, but still raises doubts about some normal condition that has been presupposed. The information about the structure of normal conditions is information that the *attributor* is presupposing, information relevant to determining what is being said about the subject's knowledge and belief. In general, the subject whose cognitive state is at issue may not even have the capacity to discriminate situations in which certain of those conditions obtain and situations in which they do not. But still, the proposition that certain background conditions hold is a factual proposition about the subject's environment and cognitive capacities, and that person must also, in a sense, be presupposing the truth of the propositions that form the background relative to which knowledge is properly attributed. At least in the case of complex knowers such as ourselves, the subject may be capable of entertaining some of these presuppositions, and may receive evidence that, while it does not *refute* any of them, still calls them into question. If Bert himself calls into question the proposition that ϕ, then that proposition cannot be one of the background conditions that is properly presupposed by an attribution of knowledge to Bert. Suppose an attributor of knowledge to Bert is presupposing a certain background normal condition ϕ, and that Bert too is initially taking ϕ for granted, but then receives evidence that this condition *may* not hold. A shift will be required to a context in which the proposition that ϕ discriminates between the relevant alternatives. I will conclude by looking at a puzzle that, as I will interpret it, suggests that a shift of this kind takes place, and that provides at least an apparent violation of the conservative principle. What I say won't resolve the puzzle, which requires further discussion, but I hope the normal-conditions framework, and the distinction between knowledge claims true in all relevant alternatives and those that discriminate between the relevant alternatives will help to clarify what is going on in the puzzle.

The puzzle is a version of an old and familiar lottery puzzle. It seems that if we want to avoid skepticism about the future, we must assume that it is possible to know certain facts about the future even when there is a small—perhaps miniscule—objective chance that they are false. It seems, for example, that I can know that a glass I have just dropped will hit the floor, even if quantum theory or statistical mechanics allows that an astronomically improbable confluence of conditions might permit it to rise up to the ceiling. This kind of puzzle has been given a particularly sharp form, using a sorites-style argument, in a recent paper by Cian Dorr, Jeremy Goodman, and John Hawthorne (DGH).[32] The setup involves a sequence of 1,000 fair coins that will be flipped in order until either one of them lands heads, or all 1,000 have been flipped. The premise of the puzzle is that Alice, who knows that the setup is

[32] Dorr, Goodman, & Hawthorne 2014. See also Salow & Goodman 2018.

as described, also knows (before the experiment begins) that the last coin in the sequence will not be flipped. But for each of the coins early in the sequence, for all she knows that coin *will* be flipped. There must, therefore, be a last coin (call it the nth flip) in the sequence such that for all Alice knows that that coin will be flipped, and this seems to have a number of implausible consequences, including this: she knows that if the nth coin is flipped, it will land tails, since otherwise it would not be the last coin flipped. But it can't be that one knows in advance how a fair coin will land if flipped.

One might question the assumption that Alice can know that the last coin won't be flipped, saying that all she knows is that that occurrence would be extremely improbable, but DGH argue that this response will not be plausible for less artificial, if messier, examples. In any case, the following seems difficult to deny: Whatever Alice's basis for initially taking herself to know that the setup is as described, if she were to learn that all 1,000 coins were flipped, she would have sufficient reason to conclude that she was mistaken about the setup. (If one chance in 2^{999} of this result, on the hypothesis that the setup is as described, is not small enough to reject the hypothesis, just increase the number of coins.) All empirical hypotheses are subject to defeat, and it would be excessively dogmatic to continue holding onto this hypothesis in the face of ever-growing evidence against it. One of the epistemic virtues of statistical methods with random trials is that they make it possible to gather evidence of increasing strength, without limit. If it is ever possible to acquire knowledge on the basis of statistical evidence, then it should be possible to come to know, from this extreme evidence, that the actual setup is not as described, and this implies that one can know, on the hypothesis that the setup *is* as described that the result (999 tails in a row) will not occur.

It cannot be denied that if the setup is as described, then there is a non-zero chance of 999 tails in a row, the same chance that any other specific sequence of heads and tails would have if the coins were all permitted to be flipped. So how can Alice know that there won't be 999 tails in a row? My suggestion, assuming a normal-conditions account of knowledge, is that there is a normality condition that applies in epistemic situations involving statistical reasoning that we might label "No misleading coincidences." (NMC). I won't try to make this precise, but the idea is that a misleading coincidence is a pattern of events that would be highly probable on some false but plausible hypotheses in the vicinity, but highly improbable on the true hypothesis. We do, I think, have some intuitive grasp of a notion like this, even without thinking of any particular alternative hypotheses. Without a theory of why this is true, I think most people would say that a sequence of twenty flips of a fair coin, all T would be much more surprising than, for example, the particular sequence, HHTHTTHHTHTHHTTTHTHH, even though these two particular alternatives are equally probable.

The rule NMC does not of course say that misleading coincidences don't ever happen, but just that they count as deviations from a norm. The attributor of knowledge to Alice may presuppose that such a condition holds, and if it does in fact hold, then Alice will know, in that context, that if the coins are fair, they will not all be flipped. But suppose the setup is as described, with the coins all fair, and the event with the prior miniscule chance of one in 2^{999} of happening does in fact occur, so that all the coins are flipped. In this case, Alice will be (initially) in a Gettier situation, with a justified true belief, but not knowledge, that the coins are all fair and flipped as described in the setup.

As with many normal conditions, NMC is a matter of degree. Ten tails in a row is pretty abnormal, but twenty much more so. The threshold for knowledge will presumably be vague, but this is not the problem. The problem is that if we were to get an increasing sequence of tails, there would inevitably come a point where the NMC presupposition is called into question before it is definitively overturned. For example, if the threshold were n, then if we got close to n, say n-2, we would (assuming the conservative principle) still take ourselves to know that the coins are fair, but we will also take ourselves to know that *if* the coins are fair, then there is a substantial chance (1 in 4) that the normality condition is violated. The normality condition (something taken to be compatible with all of the initial relevant alternatives, so known, but not known*) will then be called into question, which changes the context in a way that undercuts the claim to know that the coins are fair.

As I said, I don't have a solution to the puzzle, but I think a solution, in the context of this kind of account of knowledge, will depend on recognizing the distinction between propositions true in all relevant alternatives and propositions that discriminate between relevant alternatives, and on the way belief revision and context change interact.[33]

[33] Thanks to Bernhard Salow and Dan Greco for very helpful comments on an earlier draft of this paper, and for discussion of the issues.

PART II

Conditionals

9

A Theory of Conditionals

1. Introduction

A conditional sentence expresses a proposition that is a function of two other propositions, yet not one that is a *truth* function of those propositions. I may know the truth values of "Willie Mays played in the American League" and "Willie Mays hit four hundred" without knowing whether or not Mays would have hit four hundred if he had played in the American League. This fact has tended to puzzle, displease, or delight philosophers, and many have felt that it is a fact that calls for some comment or explanation. It has given rise to a number of philosophical problems; I will discuss three of these.[1]

My principal concern will be with what has been called the *logical problem of conditionals*, a problem that is frequently ignored or dismissed by writers on conditionals and counterfactuals. This is the task of describing the formal properties of the *conditional function:* a function, usually represented in English by the words "if... then," taking ordered pairs of propositions into propositions. I will explain informally and defend a solution, presented more rigorously elsewhere, to this problem.[2]

The second issue—the one that has dominated recent discussions of contrary-to-fact conditionals—is the *pragmatic problem of counterfactuals.* This problem derives from the belief, which I share with most philosophers writing about this topic, that the formal properties of the conditional function, together with all of the *facts,* may not be sufficient for determining the truth value of a counterfactual; that is, different truth valuations of conditional statements may be consistent with a single valuation of all non-conditional statements. The task set by the problem is to find and defend criteria for choosing among these different valuations. This problem is different from the first issue because these criteria are pragmatic, and not semantic. The distinction between semantic and pragmatic criteria, however, depends on the construction of a semantic theory. The semantic theory that I shall defend will thus help to clarify the second problem by charting the boundary between the semantic and pragmatic components of the concept. The question of this boundary line is precisely what Rescher, for example, avoids by couching his whole discussion in terms of conditions for belief, or justified belief, rather than truth conditions.[3] Conditions for justified belief are pragmatic for any concept.

[1] This chapter, first published fifty years ago, has been lightly edited for style and clarity. I have left the content unchanged, but added several footnotes in addition to this one.

[2] Stalnaker & Thomason 1970. In this paper, the formal system, C2, is proved sound and semantically complete with respect to the interpretation sketched in the present paper. That is, it is shown that a formula is a consequence of a class of formulas if and only if it is derivable from the class in the formal system, C2.

[3] Rescher 1964.

The third issue is an epistemological problem that has bothered empiricist philosophers. It is based on the fact that many counterfactuals seem to be synthetic and contingent statements about unrealized possibilities. But contingent statements must be capable of confirmation by empirical evidence, and the investigator can gather evidence only in the actual world. How are conditionals that are both empirical and contrary-to-fact possible at all? How do we learn about possible worlds, and where are the facts (or counterfacts) that make counterfactuals true? Such questions have led philosophers to try to analyze the conditional in non-conditional terms[4]—to show that conditionals merely appear to be about unrealized possibilities. My approach, however, will be to accept the appearance as reality, and to argue that one can sometimes have evidence about non-actual situations.

In sections 2 and 3 of this chapter, I will present and defend a theory of conditionals that has two parts: a formal system with a primitive conditional connective, and a semantic apparatus that provides general truth conditions for statements involving that connective. In sections 4, 5, and 6, I will discuss in a general way the relation of the theory to the three problems outlined above.

2. The Interpretation

Eventually, I want to defend a hypothesis about the truth conditions for statements having conditional form, but I will begin by asking a more practical question: How does one evaluate a conditional statement? How does someone decide whether or not he or she believes it to be true? An answer to this question will not be a set of truth conditions, but it will serve as a heuristic aid in the search for such a set.

To make the question more concrete, consider the following situation: you are faced with a true-false political opinion survey.[5] The statement is: "If the Chinese enter the Vietnam conflict, the United States will use nuclear weapons." How do you deliberate in choosing your response? What considerations of a logical sort are relevant? I shall first discuss two familiar answers to this question, and then defend a third answer that avoids some of the weaknesses of the first two.

[4] Cf. Chisholm 1946. The problem is sometimes posed (as it is here) as the task of analyzing the *subjunctive* conditional into an indicative statement, but I think it is a mistake to base very much on the distinction of mood. As far as I can tell, the mood tends to indicate something about the attitude of the speaker, but in no way affects the propositional content of the statement.

[5] These motivating remarks are indebted to Saul Kripke, who used the device of a true-false opinion survey, in the discussion following a colloquium talk, to give a counterexample to a defense of the material conditional analysis of the indicative conditional. The talk (given in 1965) was by James Thomson, and Thomson acknowledged in discussion that the counterexample was decisive, which is probably why his paper was not published in his lifetime. (The paper was published years later: Thomson 1990.) Thomson argued, using Gricean considerations, that conditionals are true when both antecedent and consequent are true even when the two propositions are unrelated, and I thought these arguments were entirely persuasive. But I also agreed that Kripke's counterexample, which concerned a conditional with a false antecedent, was decisive against the material conditional analysis, which of course cannot work for counterfactuals in any case. So, when I turned my attention to the problem, I was looking for an analysis that was stronger than the truth-functional analysis, but that could still deliver the result that conditionals could be true even when the antecedent and consequent were unrelated. [Note added in 2018]

The first answer is based on the simplest account of the conditional: the truth-functional analysis. According to this account, you should reason as follows in responding to the true-false quiz: you ask yourself, first, will the Chinese enter the conflict? Second, will the United States use nuclear weapons? If the answer to the first question is no, *or* if the answer to the second is yes, then you should place your *X* in the "true" box. But this account is unacceptable since the following piece of reasoning is an obvious *non sequitur:* "I firmly believe that the Chinese will stay out of the conflict; *therefore*, I believe that the statement is true." The falsity of the antecedent is never sufficient reason to affirm a conditional, even an indicative conditional.

A second answer is suggested by the shortcomings of the truth-functional account. The material implication analysis fails, critics have said, because it leaves out the idea of *connection* that is implicit in an if-then statement. According to this line of thought, a conditional is to be understood as a statement that affirms that some sort of logical or causal connection holds between the antecedent and the consequent. In responding to the true-false quiz, then, you should look, not at the truth-values of the two clauses, but at the relation between the propositions expressed by them. If the "connection" holds, you check the "true" box. If not, you answer "false."

If the second hypothesis were accepted, then we would face the task of clarifying the idea of "connection," but there are counter-examples even with this notion left as obscure as it is. Consider the following case: you firmly believe that the use of nuclear weapons by the United States in this war is inevitable because of the arrogance of power, the bellicosity of our president, growing pressure from congressional hawks, or other *domestic* causes. You have no opinion about future Chinese actions, but you do not think they will make much difference one way or another to nuclear escalation. Clearly, you believe the opinion survey statement to be true even though you believe the antecedent and consequent to be logically and causally independent of each other. It seems that the presence of a "connection" is not a necessary condition for the truth of an if-then statement.

The third answer I shall consider is based on a suggestion made some time ago by F. P. Ramsey.[6] Consider first the case where you have no opinion about the statement, "The Chinese will enter the Vietnam war." According to your suggestion, your deliberation about the survey statement should consist of a simple thought experiment: add the antecedent (hypothetically) to your stock of knowledge (or beliefs), and then consider whether or not the consequent is true. Your belief about the conditional should be the same as your hypothetical belief, under this condition, about the consequent.

What happens to the idea of connection on this hypothesis? It is sometimes relevant to the evaluation of a conditional, and sometimes not. If you believe that a causal or logical connection exists, then you will add the consequent to your stock of beliefs along with the antecedent, since the rational man accepts the consequences of his beliefs. On the other hand, if you already believe the consequent (and if you also believe it to be causally independent of the antecedent), then it will remain a part of

[6] Ramsey 1929. The suggestion is made on p. 248. Chisholm quotes the suggestion and discusses the limitations of the 'connection' thesis that it brings out (Chisholm 1946), but he develops it somewhat differently.

your stock of beliefs when you add the antecedent, since the rational man does not change his beliefs without reason. In either case, you will affirm the conditional. Thus, this answer accounts for the relevance of "connection" when it is relevant, without making it a necessary condition of the truth of a conditional.

Ramsey's suggestion covers only the situation in which you have no opinion about the truth-value of the antecedent. Can it be generalized? We can of course extend it without problem to the case where you believe or know the antecedent to be true; in this case, no changes need be made in your stock of beliefs. If you already believe that the Chinese will enter the Vietnam conflict, then your belief about the conditional will be just the same as your belief about the statement that the United States will use the bomb.

What about the case in which you know or believe the antecedent to be false? In this situation, you cannot simply add it to your stock of beliefs without introducing a contradiction. You must make adjustments by deleting or changing those beliefs which conflict with the antecedent. Here, the familiar difficulties begin, of course, because there will be more than one way to make the required adjustments.[7] These difficulties point to the pragmatic problem of counterfactuals, but if we set them aside for a moment, we will see a rough but general answer to the question we are asking. This is how to evaluate a conditional.

First, add the antecedent (hypothetically) to your stock of beliefs; second, make whatever adjustments are required to maintain consistency (without modifying the hypothetical belief in the antecedent); finally, consider whether or not the consequent is then true. It is not particularly important that our answer is approximate—that it skirts the problem of adjustments—since we are using it only as a way of finding truth conditions. It is crucial, however, that the answer should not be restricted to some particular context of belief if it is to be helpful in finding a definition of the conditional function. If the conditional is to be understood as a function of the propositions expressed by its component clauses, then its truth-value should not in general be dependent on the attitudes that anyone has toward those propositions.

Now that we have found an answer to the question "How do we decide whether or not we believe a conditional statement?", the problem is to make the transition from belief conditions to truth conditions; that is, to find a set of truth conditions for statements having conditional form which explains why we use the method we do use to evaluate them. The concept of a *possible world* is just what we need to make this transition, since a possible world is the ontological analogue of a stock of hypothetical beliefs. The following set of truth conditions, using this notion, is a first approximation to the account that I shall propose:

> Consider a possible world in which *A* is true, and which otherwise differs minimally from the actual world. *"If A, then B" is true (false) just in case B is true (false) in that possible world.*

[7] Rescher 1964, 11–16, contains a very clear statement and discussion of this problem, which he calls the problem of the ambiguity of belief-contravening hypotheses. He argues that the resolution of this ambiguity depends on pragmatic considerations. Cf. also Goodman's problem of relevant conditions, Goodman 1983, 17–24.

An analysis in terms of possible worlds also has the advantage of providing a ready-made apparatus on which to build a formal semantic theory. In making this account of the conditional precise, we use the semantic models for modal logics developed by Saul Kripke.[8] Following Kripke, we first define a *model structure*. Let *M* be an ordered triple $\langle K,R,\lambda\rangle$. *K* is to be understood intuitively as the set of all possible worlds; *R* is the relation of relative possibility that defines the structure. If α and β are possible worlds (members of *K*), then αRβ reads, "β is possible with respect to α." This means that, where α is the actual world, β is a possible world. *R* is a reflexive relation; that is, every world is possible with respect to itself. If your modal intuitions so incline you, you may add that *R* must be transitive, or transitive and symmetrical.[9] The only element that is not a part of the standard modal semantics is λ, a member of *K* that is to be understood as the *absurd world*—the world in which contradictions and all their consequences are true. It is an isolated element under *R;* that is, no other world is possible with respect to it, and it is not possible with respect to any other world. The purpose of A is to allow for an interpretation of "If *A*, then *B*" in the case where *A* is impossible; for this situation one needs an impossible world.[10]

In addition to a model structure, our semantic apparatus includes a *selection function, f,* which takes a proposition and a possible world as arguments and a possible world as its value. The s-function selects, for each antecedent *A*, a particular possible world in which *A* is true. The assertion that the conditional makes, then, is that the consequent is true in the world selected. A conditional is true in the actual world when its consequent is true in the selected world.

Now we can state the semantic rule for the conditional more formally (using the corner, >, as the conditional connective):

$A > B$ is true in a if B *is* true in $f(A,a)$;
$A > B$ *is* false in a if B *is* false in $f(A,a)$.

The interpretation shows conditional logic to be an extension of modal logic. Modal logic provides a way of talking about what is true in the actual world, in all possible worlds, or in at least one, unspecified world. The addition of the selection function to the semantics and the conditional connective to the object language of modal logic provides a way of talking also about what is true in *particular* non-actual possible situations. This is what counterfactuals are: statements about particular counterfactual worlds.

But the world selected cannot be just any world. The s-function must meet four conditions. (I will use the following terminology for talking about the arguments and values of s-functions: where $f(A,\alpha) = \beta$, *A* is the *antecedent*, α is the *base world*, and β is the *selected world*.)

[8] Kripke 1963.

[9] The different restrictions on the relation *R* provide interpretations for the different modal systems. The system we build on is von Wright's M. If we add the transitivity requirement, then the underlying modal logic of our system is Lewis's S4. and if we add both the transitivity and symmetry requirements, then the modal logic is SS. Cf. Kripke, op. cit.

[10] The absurd world was not meant to be taken seriously. The analysis could be stated without it, simply by stipulating that the conditional is true when there is no world eligible to be selected. [Note added in 2018].

(1) For all antecedents A and base worlds α, A must be true in $f(A,\alpha)$.
(2) For all antecedents A and base worlds α, $f(A,\alpha) = \lambda$ only if there is no world possible with respect to α in which A is true.[11]

The first condition requires that the antecedent be true in the selected world. This ensures that all statements like "if snow is white, then snow is white" are true. The second condition requires that the absurd world be selected only when the antecedent is impossible. Since everything is true in the absurd world, including contradictions, if the selection function were to choose it for the antecedent *A*, then "If *A*, then *B* and not *B*" would be true. But one cannot legitimately reach an impossible conclusion from a consistent assumption.

The informal truth conditions that were suggested above required that the world selected *differ minimally* from the actual world. This implies, first, that there are no differences between the actual world and the selected world except those that are required, implicitly or explicitly, by the antecedent. Further, it means that among the alternative ways of making the required changes, one must choose one that does the least violence to the correct description and explanation of the actual world. These are vague conditions that are largely dependent on pragmatic considerations for their application. They suggest, however, that the selection is based on an ordering of possible worlds with respect to their resemblance to the base world. If this is correct, then there are two further formal constraints that must be imposed on the s-function:

(3) For all base worlds α and all antecedents A, if A is true in α then $f(A,\alpha) = \alpha$.
(4) For all base worlds a and all antecedents B and B', if B is true in $f(B',\alpha)$ and B' is true in $f(B,\alpha)$, then $f(B,\alpha) = f(B',\alpha.)$.

The third condition requires that the base world be selected if it is among the worlds in which the antecedent is true. Whatever the criteria for evaluating resemblance among possible worlds, there is obviously no other possible world as much like the base world as the base world itself. The fourth condition ensures that the ordering among possible worlds is consistent in the following sense: if any selection established β as prior to β' in the ordering (with respect to a particular base world α), then no other selection (relative to that α) may establish β' as prior to β.[12] Conditions (3) and (4) together ensure that the s-function establishes a total ordering of all

[11] Shouldn't it be "if and only if" rather than just "only if"? I had thought that the "if" part was a consequence of the other conditions, but closer examination reveals an anomaly in the semantics that went unnoticed (at least by me) until Matt Mandelkern pointed it out recently to me. The original idea was to start with a Kripke model, with its interpretation of the necessity and possibility operators, and then add some additional structure with the selection function. But then it was observed that the necessity and possibility operators could be defined in terms of the conditional, so there is no separate semantic rule for the standard modal operators. The accessibility relation becomes an idle wheel in the semantics. But we still want to show that the standard modal operators have their standard interpretation, and to do this we need to show that the Kripkean semantic rules for the standard operators could be derived from the semantic rules for the conditional connective. Condition (2), strengthened to "if and only if" and now thought of as a constraint on R rather than a constraint on the selection function, will allow us to show that the defined "□φ" is true in world x iff φ is true in all β such that αRβ. (Thanks to Matt Mandelkern for his observation, and discussion of it.) [ote added in 2018.]

[12] If $f(A,\alpha) = \beta$, then β is established as prior to all worlds possible with respect to α in which *A* is true.

selected worlds with respect to each possible world, with the base world preceding all others in the order.

These conditions on the selection function are necessary in order that this account be recognizable as an explication of the conditional, but they are of course far from sufficient to determine the function uniquely. There may be further formal constraints that can plausibly be imposed on the selection principle, but we should not expect to find semantic conditions sufficient to guarantee that there will be a unique s-function for each valuation of non-conditional formulas on a model structure. The questions, "On what basis do we select a selection function from among the acceptable ones?" and "'What are the criteria for ordering possible worlds?" are reformulations of the pragmatic problem of counterfactuals, which is a problem in the application of conditional logic. The conditions that I have mentioned above are sufficient, however, to define the semantic notions of validity and consequence for conditional logic.

3. The Formal System

The class of valid formulas of conditional logic, according to the definitions sketched in the preceding section, is coextensive with the class of theorems of a formal system, C2. The primitive connectives of C2 are the usual "$\supset$" and "~"{with "$\vee$," "&," and "$\equiv$" defined as usual), as well as a conditional connective, ">" (called the corner). The necessity and possibility operators, and a bi-conditional operator can be defined in terms of the corner as follows:

$\Box A =_{df} (\sim A > A)$
$\Diamond A =_{df} \sim(A > \sim A)$
$(A \asymp B) =_{df} (A > B) \& (B > A)$

The rules of inference of C2 are *modus ponens* (if A and $A \supset B$ are theorems, then B is a theorem) and the rule of necessitation (If A is a theorem, then $\Box A$ is a theorem). There are seven axiom schemata:

(al) Any tautologous sentence is an axiom
(a2) $\Box(A \supset B) \supset (\Box A \supset \Box B)$
(a3) $\Box(A \supset B) \supset (A > B)$
(a4) $\Diamond A \supset ((A > B) \supset \sim(A > \sim B))$
(a5) $(A > (B \vee C)) \supset ((A > B) \vee (A > C))$
(a6) $(A > B) \supset (A \supset B)$
(a7) $((A \asymp B) \supset ((A > C) \supset (B > C))$

The conditional connective, as characterized by this formal system, is intermediate between strict implication and the material conditional in the sense that $\Box(A \supset B)$ entails $(A > B)$ (by (a3)) and $(A > B)$ entails $(A \supset B)$ (by a6). It cannot, however, be analyzed as a modal operation performed on a material conditional (like Burks's causal implication, for example).[13] The corner lacks certain properties shared by the

[13] Burks 1951. The causal implication connective characterized in this article has the same structure as strict implication. For an interesting philosophical defense of this modal interpretation of conditionals, see Mayo 1957.

two traditional implication concepts, and in fact these differences help to explain some peculiarities of counterfactuals. I shall point out three unusual features of the conditional connective.

(1) Unlike both material and strict implication, the conditional comer is a non-transitive connective. That is, from $A > B$ and $B > C$, one cannot infer $A > C$. While this may at first seem surprising, consider the following example: *Premises.* "If J. Edgar Hoover were today a Communist, then he would be a traitor." "If J. Edgar Hoover had been born a Russian, then he would today be a Communist." *Conclusion.* "If J. Edgar Hoover had been born a Russian, he would be a traitor." It seems reasonable to affirm these premises and deny the conclusion.

If this example is not sufficiently compelling, note that the following rule follows from the transitivity rule: From $A > B$ to infer $(A\&C) > B$. But it is obvious that the former rule is invalid; we cannot always strengthen the antecedent of a true conditional and have it remain true. Consider "If this match were struck, it would light," and "If this match had been soaked in water overnight *and* it were struck, it would light."[14]

(2) According to the formal system, the denial of a conditional is equivalent to a conditional with the same antecedent and opposite consequent (provided that the antecedent is not impossible). That is, $\Diamond A \supset ((A > B) \equiv {\sim}(A > {\sim}B))$. This explains the fact, noted by both Goodman and Chisholm in their early papers on counterfactuals, that the normal way to contradict a counterfactual is to contradict the consequent, keeping the same antecedent. To deny "If Kennedy were alive today, we wouldn't be in this Vietnam mess," we say, "If Kennedy were alive today, we would so be in this Vietnam mess."

(3) The inference of contraposition, valid for both the truth-functional horseshoe and the strict implication hook, is invalid for the conditional corner. $A > B$ may be true while ${\sim}B > {\sim}A$ is false. For an example in support of this conclusion, we take another item from the political opinion survey: "If the U.S. halts the bombing, then North Vietnam will not agree to negotiate." A person would believe that this statement is true if he thought that the North Vietnamese were determined to press for a complete withdrawal of US troops. But he would surely deny the contrapositive, "If North Vietnam agrees to negotiate, then the US will not have halted the bombing." He would believe that a halt in the bombing, and much more, is required to bring the North Vietnamese to the negotiating table.[15]

Examples of these anomalies have been noted by philosophers in the past. For instance, Goodman pointed out that two counterfactuals with the same antecedent and contradictory consequents are "normally meant" as direct negations of each other. He also remarked that we may sometimes assert a conditional and yet reject its contrapositive. He accounted for these facts by arguing that semifactuals

[14] Although the transitivity inference fails, a related inference is of course valid. From $A > B$, $B > C$, and A, one can infer C. Also, note that the bi-conditional connective is transitive. From $A \asymp B$ and $B \asymp C$, one can infer $A \asymp C$. Thus the bi-conditional relation is an equivalence relation since it is also symmetric and reflexive.

[15] Although contraposition fails, *modus tollens* is valid for the conditional: from $A > B$ and ${\sim}B$, one can infer ${\sim}A$.

(conditionals with false antecedents and true consequents) are for the most part not to be taken literally. "In practice," he wrote, "full counterfactuals affirm, while semifactuals deny, that a certain connection obtains between antecedent and consequent.... The practical import of a semifactual is thus different from its literal import.'[16] Chisholm also suggested paraphrasing semifactuals before analyzing them. "Even if you were to sleep all morning, you would be tired" is to be read, "It is false that if you were to sleep all morning, you would not be tired."[17]

A separate and non-conditional analysis for semifactuals is necessary to save the "connection" theory of counterfactuals in the face of the anomalies we have discussed, but it is a baldly ad hoc manoeuver. Any analysis can be saved by paraphrasing the counterexamples. The theory presented in section 2 avoids this difficulty by denying that the conditional can be said, in general, to assert a connection of any particular kind between antecedent and consequent. It is, of course, the structure of inductive relations and causal connections that makes counterfactuals and semifactuals true or false, but it does this by determining the relationships among possible worlds, which in turn determine the truth-values of conditionals. By treating the relation between connection and conditionals as an indirect relation in this way, the theory is able to give a unified account of conditionals that explains the variations in their behavior in different contexts.

4. The Logical Problem: General Considerations

The traditional strategy for attacking a problem like the logical problem of conditionals was to find an *analysis*: to show that the unclear or objectionable phrase was dispensable, or replaceable by something clear and harmless. Analysis was viewed by some as an *unpacking*—a making manifest of what was latent in the concept; by others it was seen as the *replacement* of a vague idea by a precise one, adequate to the same purposes as the old expression, but free of its problems. The semantic theory of conditionals can also be viewed either as the construction of a concept to replace an unclear notion of ordinary language, or as an *explanation* of a commonly used concept. I see the theory in the latter way: no recommendation or stipulation is intended. This does not imply, however, that the theory is meant as a description of linguistic usage. What is being explained is not the rules governing the use of an English word, but the structure of a concept. Linguistic facts—what we would say in this or that context, and what sounds odd to the native speaker—are relevant as evidence, since one may presume that concepts are to some extent mirrored in language.

The "facts," taken singly, need not be decisive. A recalcitrant counterexample may be judged a deviant use or a different sense of the word. We can claim that a paraphrase is necessary, or even that ordinary language is systematically mistaken about the concept we are explaining. There are, of course, different senses and times when "ordinary language" goes astray, but such ad hoc hypotheses and qualifications diminish both the plausibility and the explanatory force of a theory. While we are not irrevocably bound to the linguistic facts, there are no "don't cares"—contexts of use

[16] Goodman 1955, 15, 32. [17] Chisholm 1946, 492.

with which we are not concerned—since any context can be relevant as evidence for or against an analysis. A general interpretation which avoids dividing senses and accounts for the behavior of a concept in many contexts fits the familiar pattern of scientific explanation in which diverse, seemingly unlike surface phenomena are seen as deriving from some common source. For these reasons, I take it as a strong point in favor of the semantic theory that it treats the conditional as a univocal concept.

5. Pragmatic Ambiguity

I have argued that the conditional connective is semantically unambiguous. It is obvious, however, that the context of utterance, the purpose of the assertion, and the beliefs of the speaker or his community may make a difference to the interpretation of a counterfactual. How do we reconcile the ambiguity of conditional sentences with the univocality of the conditional concept? Let us look more closely at the notion of ambiguity.

A sentence is ambiguous if there is more than one proposition that it may properly be interpreted to express. Ambiguity may be syntactic (if the sentence has more than one grammatical structure), semantic (if one of the words has more than one meaning), or pragmatic (if the interpretation depends directly on the context of use). The first two kinds of ambiguity are perhaps more familiar, but the third kind is probably the most common in natural languages. Any sentence involving pronouns, tensed verbs, articles, or quantifiers is pragmatically ambiguous. For example, the proposition expressed by "L'etat, c'est moi" depends on who says it; "Do it now" may be good or bad advice depending on when it is said; "Cherchez la femme" is ambiguous since it contains a definite description; and the truth conditions for "All's well that ends well" depend on the domain of discourse. If the theory presented above is correct, then we may add conditional sentences to this list. The truth conditions for "If wishes were horses, then beggars would ride" depend on the specification of an s-function.[18]

The grounds for treating the ambiguity of conditional sentences as pragmatic rather than semantic are the same as the grounds for treating the ambiguity of quantified sentences as pragmatic: simplicity and systematic coherence. The truth-conditions for quantified statements vary with a change in the domain of discourse, but there is a single structure to these truth conditions that remains constant for every domain. The semantics for classical predicate logic brings out this common structure by giving the universal quantifier a single meaning and making the domain a parameter of the interpretation. In a similar fashion, the semantics for conditional logic brings out the common structure of the truth-conditions for conditional statements by giving the connective a single meaning and making the selection function a parameter of the interpretation.

[18] I do not wish to pretend that the notions needed to define ambiguity and to make the distinction between pragmatic and semantic ambiguity (e.g., "proposition" and "meaning") are precise. They can be made precise only in the context of semantic and pragmatic theories. But even if it is unclear, in general, what pragmatic ambiguity is, it is clear, I hope, that my examples are cases of it.

Just as we can communicate effectively using quantified sentences without explicitly specifying a domain, so we can communicate effectively using conditional sentences without explicitly specifying an s-function. This suggests that there are further rules beyond those set down in the semantics, governing the use of conditional sentences. Such rules are the subject matter of a *pragmatics* of conditionals. Very little can be said, at this point, about pragmatic rules for the use of conditionals since the logic has not advanced beyond the propositional stage, but I will make a few speculative remarks about the kind of research that may provide a framework for treatment of the problem, and related pragmatic problems in the philosophy of science.

(1) If we had a quantified conditional logic, it is likely that $(\forall x)(Fx > Gx)$ would be a plausible candidate for the form of a law of nature. A law of nature says, not just that every actual F is a *G*, but further that for every possible *F*, if it were an *F*, it would be a *G*. If this is correct, then Hempel's confirmation paradox does not arise, since "All ravens are black" is not logically equivalent *to* "All non-black things are non-ravens." Also, the relation between counterfactuals and laws becomes clear: laws support counterfactuals because they entail them. "If this dove were a raven, it would be black" is simply an instantiation of "All ravens are black."[19]

(2) Goodman argued that the pragmatic problem of counterfactuals is one of a cluster of closely related problems concerning induction and confirmation. He locates the source of these difficulties in the general problem of projectability, which can be stated roughly as follows: When can a predicate be validly projected from one set of cases to others? Or, when is a hypothesis confirmed by its positive instances? Some way of distinguishing between natural predicates and those that are artificially constructed is needed. If a theory of projection such as Goodman envisions were developed, it might find a natural place in a pragmatics of conditionals. Pragmatic criteria for measuring the inductive properties of predicates might provide pragmatic criteria for ordering possible worlds.[20]

(3) There are some striking structural parallels between conditional logic and conditional probability functions, which suggests the possibility of a connection between inductive logic and conditional logic. A probability assignment and an s-function are two quite different ways of describing the inductive relations among propositions; a theory that draws a connection between them might be illuminating for both.[21]

6. Conclusion: Empiricism and Possible Worlds

Writers of fiction and fantasy sometimes suggest that imaginary worlds have a life of their own beyond the control of their creators. Pirandello's six characters, for example,

[19] For a discussion of the relation of laws to counterfactuals, see Nagel 1961, 47–78. For a recent discussion of the paradoxes of confirmation by the person who discovered them, see Hempel 1966.

[20] Goodman 1955, especially ch. 4.

[21] Several philosophers have discussed the relation of conditional propositions to conditional probabilities. See Jeffrey 1964 and Adams 1966. I hope to present elsewhere my method of drawing the connection between the two notions, which differs from both of these.

rebelled against their author and took the story out of his hands. The skeptic may be inclined to suspect that this suggestion is itself fantasy. He believes that nothing goes into a fictional world, or a possible world, unless it is put there by decision or convention: it is a creature of invention and not discovery. Even the fabulist Tolkien admits that Faerie is a land "full of wonder, but not of information."[22]

For similar reasons, the empiricist may be uncomfortable about a theory that treats counterfactuals as literal statements about non-actual situations. Counterfactuals are often contingent, and contingent statements must be supported by evidence. But evidence can be gathered, by us at least, only in this universe. To satisfy the empiricist I must show how possible worlds, even if the product of convention, can be subjects of empirical investigation.

There is no mystery to the fact that I can partially define a possible world in such a way that I am ignorant of some of the determinate truths in that world. One way I can do this is to attribute to it features of the actual world that are unknown to me. Thus, I can say, "I am thinking of a possible world in which the population of China is just the same, on each day, as it is in the actual world." *I* am making up this world—it is a pure product of my intentions—but there are already things true in it that I will never know.

Conditionals do implicitly, and by convention, what is done explicitly by stipulation in this example. It is because counterfactuals are generally about possible worlds that are very much like the actual one, and defined in terms of it, that evidence is so often relevant to their truth. When I wonder, for example, what would have happened if I had asked my boss for a raise yesterday, I am wondering about a possible world that I have already roughly picked out. It has the same history, up to yesterday, as the actual world, the same boss with the same dispositions and habits. The main difference is that in that world, yesterday I asked the boss for a raise. Since I do not know everything about the boss's habits and dispositions in the actual world, there is a lot that I do not know about how he acts in the possible world that I have chosen, although I might find out by watching him respond to a similar request from another, or by asking his secretary about his mood yesterday. These bits of information about the actual world would not be decisive, of course, but they would be relevant, since they tell me more about the non-actual situation that I have selected.

If I make a conditional statement, subjunctive or otherwise, and the antecedent turns out to be true, then whether I know it or not, I have said something about the actual world, namely that the consequent is true in it. If the antecedent is false, then I have said something about a particular counterfactual world, even if I believe the antecedent to be true. The conditional provides a set of conventions for selecting possible situations that have a specified relation to what actually happens. This makes it possible for statements about unrealized possibilities to tell us, not just about the speaker's intentions, but about the world.

[22] Tolkien 1966, 3.

10

Conditional Propositions and Conditional Assertions

1. Introduction

One standard way of approaching the problem of analyzing conditional sentences begins with the assumption that a sentence of this kind expresses a proposition that is a function of the propositions expressed by its component parts (plus, perhaps, some features of the context in which the sentence is uttered). The task is to characterize this function. But there is also a long tradition according to which conditional sentences—at least some conditional sentences—are used to perform a special kind of speech act. A conditional assertion is not a standard kind of speech act (assertion) with a distinctive kind of content (a conditional proposition), but rather a distinctive kind of speech act that involves just the two propositions, the ones expressed by the antecedent and the consequent. There has been considerable controversy about which of these two strategies for explaining conditionals is better.

There is a second controversial issue that interacts with this one: conditional sentences have traditionally been divided into two categories, usually labeled "subjunctive" and "indicative," even though it has long been recognized that while there is a clear grammatical contrast between the two kinds of conditionals, the difference is not a simple matter of grammatical mood. The issue concerns the relationship between the conditionals of the two kinds. Some theorists have treated the problem of analyzing indicative and subjunctive conditionals as separate problems, each to be treated on its own terms. Others have sought some kind of unified analysis. It is clear that the two kinds of conditionals have much in common, but also clear that there are semantic differences between them since there are minimal pairs, differing only in that one is "subjunctive" and the other "indicative" that seem, intuitively, to say quite different things.[1] So while it is uncontroversial that the contrast between the two kinds of conditionals is not a simple and superficial one, there remains a question whether one can explain the semantic and pragmatic differences within a unified theory of conditionals, or whether one should treat the two kinds of "ifs" as different concepts that happen to be expressed by the same word. The issue about whether

[1] By a *semantic* difference, I mean here a difference in the assertive content that utterances of the contrasting conditional sentences would have in a similar situation. This is compatible with the hypothesis that the abstract semantics for the two conditionals is the same, but that the difference in content is explained by a difference in contextual determinants relative to which the contrasting kinds of conditionals are interpreted.

conditionals express propositions interacts with the question whether we can give a unified account of the two kinds of conditionals, since the considerations favoring the propositional analysis of conditionals are much stronger in the case of subjunctive conditionals, while the considerations favoring the conditional assertion account are much stronger in the case of indicative conditionals.

Some philosophers—David Lewis and Frank Jackson, for example—make no attempt to give a unified account that covers both kinds of conditionals, but still support the hypothesis that indicative conditionals have truth conditions. Lewis gave the following reason for this decision:

> I have no conclusive objection to the hypothesis that indicative conditionals are non-truth-valued sentences, governed by a special rule of assertability. . . . I have an inconclusive objection, however: the hypothesis requires too much of a fresh start. It burdens us with too much work still to be done, and wastes too much that has been done already. . . . We think we know how the truth conditions for compound sentences of various kinds are determined by the truth conditions of constituent subsentences, but this knowledge would be useless if any of those subsentences lacked truth conditions. Either we need new semantic rules for many familiar connectives and operators when applied to indicative conditionals—perhaps rules of truth, perhaps special rules of assertability like the rule for conditionals themselves—or else we need to explain away all seeming examples of compound sentences with conditional constituents.[2]

Lewis's methodological concern might be generalized. It is not only that if we treat indicative conditionals as truth-conditional, we can draw on the resources of compositional semantics to explain the embedding of conditionals in other constructions, it is also that we can draw on standing accounts of speech acts, such as assertion, and of propositional attitudes such as belief, and epistemic states such as knowledge to explain the assertion of, belief in, and knowledge of conditionals. Speech acts and propositional attitudes are standardly factored into content and force, or content and kind of attitude. By treating a conditional as a distinctive kind of content, one avoids the problem of giving an account of distinctive kinds of conditional force, and distinctive conditional mental states.

But these methodological considerations cut both ways. While there are some complex constructions with indicative conditionals as constituents, the embedding possibilities seem, intuitively, to be highly constrained. For example, simple disjunctions of indicative conditionals with different antecedents, and conditionals with conditional antecedents are sometimes difficult to make sense of. The proponent of a non-truth-conditional account needs to explain what embeddings there are, but the proponent of a truth-conditional account must explain why embedded conditionals don't seem to be interpretable in full generality. And while the truth-functional analysis of indicative conditionals favored by Lewis, Jackson, and Paul Grice can make sense of negations and disjunctions with conditionals parts, it is in those kinds of constructions that the consequences of the truth-functional analysis are most difficult to reconcile with intuitions about examples.

[2] Lewis 1976.

On the speech act and attitude issue: it would be an advantage to reduce the problem of conditional assertion, promise, command, belief, intention, knowledge, and so forth to the single problem of analyzing a kind of content—conditional propositions—that they all share, but on the other hand, it may distort the phenomenon of conditionality to do so. As is often noted, conditional bets and questions are not properly understood as bets on or questions about the truth of a conditional proposition, and conditional promises do not seem to be promises to make true a conditional proposition. It may also be that a proper account of belief, intention, and knowledge must make room for conditional versions of these attitudes that cannot be reduced to categorical belief, knowledge, and intention.

So, there are costs and benefits on both sides to be weighed in comparing the truth-conditional and non-truth-conditional accounts of conditionals. My aim in this chapter is to try to get clearer about what is at stake in this debate, focusing on the case of indicative conditional assertions. More generally, I hope to get clearer about the relation between speech acts and the propositions and propositional attitudes that are expressed in them. My strategy will be to sketch a specific account of each kind within a common framework, and then to consider exactly how they differ. One of my conclusions will be that while there are real differences between the accounts, they may be less significant than they have seemed.

In section 2, I will sketch a conditional assertion account, which will require saying something about how categorical assertion should be understood. In section 3, I will sketch a non-truth functional propositional account of indicative conditionals, an account that is basically the same as one I proposed some years ago. While this account has problems, it is not so easily refuted as has been suggested. I will respond to some arguments against it, but also say what I think the real problems with the account are. There are tensions, I will suggest, between the roles of public knowledge (or common ground) and the knowledge and beliefs of individual speakers in determining what is said in indicative conditional statements. I will conclude, in section 4, with an example that brings out this tension, and that is a problem for both propositional and conditional assertion accounts.

2. Conditional Speech Acts

W. V. Quine, in an often-quoted remark, said that

> An affirmation of the form "if p then q" is commonly felt less as an affirmation of a conditional than as a conditional affirmation of the consequent. If, after we have made such an affirmation, the antecedent turns out true, then we consider ourselves committed to the consequent, and are ready to acknowledge error if it proves false. If on the other hand the antecedent turns out to have been false, our conditional affirmation is as if it had never been made.[3]

G. H. von Wright, in a 1957 article on conditionals, proposed to treat the conditional as a "mode of asserting," and tentatively suggested that conditional sentences do not express propositions: "I shall never speak of a conditional as a proposition which is

[3] Quine 1959, 12.

being asserted, but only of propositions being asserted conditionally, relative to other propositions."[4] J. Mackie characterized a conditional assertion, if p, then q, as an assertion of q within the scope of the supposition that p.[5] But how exactly is the speech act of conditional assertion to be understood?

To better understand what needs to be said to answer this question, it will help to say a little bit about how the speech act of categorical assertion is to be understood. There are two ways of approaching the task of giving an account of a speech act such as assertion, both of which have their roots in J. L. Austin's work on speech acts. Speech acts obviously alter the situation in which they take place, and one might try to explain what it is to make an assertion by saying how it changes, or is intended to change, the context. Alternatively, one might characterize assertions in terms of the way they are assessed. Speech acts are generally assumed to be moves in a rule-governed institutional practice, and one might focus on the constitutive norms that constrain the practice. A speech act might be successful in the sense that it succeeds in changing the context in the way that assertions are intended to change the context, but still be defective in some way—still be an assertion that failed to meet some standard or norm that assertions are supposed to meet. A full account of assertion should include an account of such standards or norms.

David Lewis, in his "Score keeping in a language game," sketches a framework for answering the first question. He suggested that we think of a conversation as like a game with an evolving score, and of speech acts as moves in the game. A characterization of a speech act of a certain kind will be an account of how the speech act changes the score of the game. My account of assertion in "Assertion"[6] fit this pattern. A discourse context is represented by a set of possible situations—a context set—representing the relevant alternatives, or live options that the conversational participants intend to distinguish between in their speech acts. The essential effect of an assertion is to add the content of the assertion to the information that is henceforth to be presupposed—to eliminate from the context set those possible situations that are incompatible with the content of the assertion. On this account, one might think of an assertion as something like a proposal to change the context set in that way, a proposal that is adopted if it is not rejected by one of the other parties to the conversation.

Different theorists might accept this account of the effect of assertions, but give different answers to the question about how assertions are assessed, and one might distinguish different speech acts in terms of the way they are assessed, even if they change the context in the same way. For example, parties to a conversation might agree to accept certain things that they may not be in a position to assert. ("It's probably going to rain—let's assume that it will." Or lawyers in a court proceeding may agree to stipulate certain facts.) This kind of speech act changes the context in the same way as an assertion, but may be subject to different norms.[7]

[4] von Wright 1957, 131. [5] Mackie 1973.

[6] Chapter 4 of Stalnaker 1999a, originally published in 1978.

[7] I don't want to suggest that there is a sharp or deep line between the two kinds of questions about speech acts (how they change the context, and how they are assessed). If two speech acts are subject to different norms, then they will inevitably change the context in different ways as a result. For example, if

Here are some contrasting answers to the question about the norms of assertion that have been given:

(1) Some claim that the only norm for assertion is truth. That is, an assertion of something false is defective, subject to criticism, but a true assertion achieves all that assertions essentially aim at. This account need not say that a speaker who makes a false assertion is always subject to criticism—she might have had good reason to believe that her action conformed to the norm. And conversely, the account will allow that speakers may be subject to criticism even for true assertions, if they did not have good enough reason to believe that they were true. There are general norms that one should take care, in one's actions, to ensure meet the standards that apply to those actions. The claim is that the only specific norm governing assertion is the norm of truth.[8]

(2) Others—most prominently, Timothy Williamson—have argued that successful assertions must meet a higher standard: the speaker represents himself as knowing the content of the assertion, and so the assertion fails of its aim if the speaker does not have this knowledge.[9] Again, speakers who make assertions without knowledge may not be subject to criticism, should they be justified in believing that they knew the truth of what they asserted, but their assertions will still be defective when the speaker lacks knowledge.[10]

(3) Robert Brandom and others have argued that one who makes an assertion undertakes a commitment to defend the truth of the assertion in response to reasonable challenges—a proposed norm that is weaker in some respects, and stronger in others than the norm of knowledge.[11] Thus, an assertion by one who is uncertain of the truth of what he asserts, but who is prepared to give arguments, and to withdraw the claim in the face of good counter arguments satisfies this norm, while failing to satisfy the knowledge norm. On the other hand, one who says, "I will tell you what I know, but you will just have to take my word for it" may satisfy the knowledge norm, while failing to live up to this one.

(4) A Bayesian might say that assertability is a matter of degree.[12] The higher the degree of belief that the speaker has in the truth of the content of the assertion, the more asssertable it is. How high is high enough will depend on context—on the balance of the costs of being wrong, and the benefits of getting it right.

knowledge is a mutually recognized norm of assertion, then an assertion that *P* will normally change the context by making it common ground that the speaker has represented herself as knowing that *P*. But different theorists have emphasized one or the other of these questions, and I think it is useful to think of their answers as complementary parts of a full account of a speech act.

[8] See Weiner 2005 for a defense of this thesis.

[9] Williamson 1996 and 2000, chapter 11.

[10] The difference between the truth norm and the knowledge norm is subtle, and some might argue that there is really no difference, given that to meet the norm of truth (together with the general norm that one should have sufficient reason to believe that one is conforming to any specific norm), one must have sufficient reason to believe that one's assertion is true, and that might be hard to distinguish from sufficient reason to believe that one had knowledge. But Williamson argues that there is a difference, and that it is important.

[11] Brandom 1983.

[12] I don't mean to suggest that anyone who accepts a Bayesian account of belief and degree of belief is committed to any particular line on norms of assertion.

Now with those models of what an account of a speech act might look like, how might one explain conditional assertion? First, here is an answer to the first question, paralleling the account of the effect of categorical assertion on the context: First, one adds the content of the antecedent, temporarily, to the context; that is, one sets aside the possibilities in the context set in which the supposition is false.[13] (What if the supposition is incompatible with the prior context, so that the whole context set is set aside? In this case, just as in the case of a categorical assertion that is incompatible with the prior context, the speech act will be inappropriate unless there is a way to adjust the context to make it compatible with the supposition.) Then the content of the consequent is treated like the content of a categorical assertion: one eliminates, from this temporary or derived context those possible situations that are incompatible with the content of the consequent. Finally, one adds back the possibilities that one had set aside.

David Lewis, and other defenders of the material conditional analysis, might point out that on this account, the effect of a conditional assertion is exactly the same as the effect of the categorical assertion of the corresponding material conditional. Is this just an example of Lewis's methodological point that a non-truth conditional account will have to do over again work that has already been done? The defender of the conditional assertion analysis will argue that it is not. Even if the effect of a conditional assertion is exactly the same as the effect of a categorical assertion of the corresponding material conditional, that does not mean that the overall account of conditionals is the same on the two accounts. It was, after all, Lewis's point that the material conditional analysis yields an account of the role of conditionals in embedded contexts, while the conditional assertion account (without further supplementation) does not. And it is with embedded conditionals that the material conditional account runs into trouble.

The most striking counterexamples to the material conditional analysis of conditionals are negations of conditionals. My favorite is an argument for the existence of God cited by Dorothy Edgington, and attributed to W. D. Hart: If there is no god, then it is not the case that if I pray, my prayers will be answered. I don't pray; therefore, there is a god. The premisses seem more reasonable than the conclusion, but if the conditional is the material conditional, the argument is valid.[14] On the conditional assertion account, as Lewis observed, there is no straightforward interpretation of the negation of an indicative conditional, but it is natural to interpret it as a conditional denial. To deny that if I pray, my prayers will be answered is the same as to assert that my prayers will not be answered, conditional on my praying. (In the argument, the negation of the conditional is also embedded in a conditional context, but the conditional assertion account has no problem with conditionals with conditional consequents. "If *A*, then if *B*, then *C*" is a conditional assertion of *C* on condition *B*, made in the context of a supposition that *A*.) The defender of the material conditional analysis may try to explain the discrepancy between intuitive judgments about negations of conditionals and what is implied by his analysis by

[13] Swanson 2004 argues that we should think of conditional assertions as two different speech acts—an act of supposition followed by an assertion under the scope of the supposition.

[14] Edgington 1986, 187.

paraphrasing the negative conditional, putting the negation on the consequent (perhaps using some kind of general bracketing device, such as Grice proposed in his defense of the material conditional analysis[15]), but to do this is to give up the methodological advantage that Lewis claimed for a truth-conditional account.

Negations of indicative conditionals often seem interpretable, and when they are, are most naturally interpreted as conditionals with the negation on the consequent. In contrast, disjunctions of indicative conditionals (with different antecedents) are hard to make sense of. Suppose I were to say, "Either he will win if he carries Ohio, or he will carry Ohio if he wins." If you can make sense of this at all, does it seem to be a tautology (as it is, on the material conditional analysis)? On the other hand, *sometimes* disjunctions of indicative conditionals seem fine, especially when they are future oriented, where the semantic difference between indicative and subjunctive is subtler. Suppose there are two switches, one of which controls the light, but I don't know which it is. I say, "If I flip switch A, the light will go on, or if I flip switch B, the light will go on." The syntax may be a bit awkward, but the meaning seems clear enough.[16]

The conditional assertion account yields no natural interpretation for a disjunction of conditionals with different antecedents. (If the antecedents of the disjoined conditionals are the same, then they might naturally be interpreted as an assertion of the disjunction of the consequents, conditional on their common antecedent.) If you find such conditionals intuitively bewildering, this is a point (at least a small point) in favor of the conditional assertion account, but it is a point against that you need to explain the cases that seem okay.

So, the conditional assertion account yields the result that the effect of a conditional assertion is the same as the effect of a categorical assertion of a material conditional, but this does not imply that the two accounts are equivalent for just the reason that Lewis emphasized: if conditionals have truth conditions they may be embedded in other truth-conditional contexts. The two accounts may also differ in what they say about how conditional assertions should be assessed. What are the norms of conditional assertion? Each of the alterative accounts of the norms of categorical assertion suggests a natural extension to an account of the norms of conditional assertion. Let me consider them in turn.

(1) If truth is the sole norm of assertion, then the natural extension to conditional assertion would be to say that a conditional assertion is subject to a conditional norm of truth (of the consequent), conditional on the truth of the antecedent. This is the norm suggested by the remark from Quine quoted above. The sense in which the conditional affirmation is "as if it had never been made" in the case where the antecedent is false is that it is then not subject to criticism for violating the norm. On this account, the norm for conditional assertion is, in effect, the same as the norm for a categorical assertion of a material conditional.[17]

[15] Grice 1989, Ch. 4.

[16] This point was made by an anonymous referee, with this example.

[17] There will always be general Gricean norms of cooperative speech, constraining both categorical and conditional assertions, but these will not be norms that are specific to a kind of speech act, but general constraints on rational cooperative behavior. As Grice noted, it is normally misleading to assert a (material) conditional on the basis of knowledge of the falsity of the antecedent, and he offered an

(2) If knowledge is the norm for categorical assertion, then it seems reasonable to require that a speaker have conditional knowledge of the consequent, conditional on the truth of the antecedent. But what is conditional knowledge? In the case where a person is unsure whether a condition *A* is true or false, then it seems reasonable to say that she has knowledge of *B*, conditional on *A*, just in case she has categorical knowledge of the material conditional ($A \supset B$), so at least in the case where the speaker is uncertain about the truth of the antecedent, the norm will be, in effect, the same as the norm for the categorical assertion of the corresponding material conditional. But what about situations where the speaker knows the antecedent to be false? In this case, she knows the truth of the material conditional, but if we can make sense of conditional knowledge in this kind of situation, then there might be cases where one would satisfy the norm for the categorical assertion of the material conditional, while violating the norm for conditional assertion.

One might be tempted to think that a conditional assertion, made by a speaker who knows that the antecedent is false, would always violate some Gricean maxims, since such a speaker is always in a position to assert that the antecedent is false, and if the falsity of the antecedent were asserted and accepted, the conditional assertion would no longer be appropriate (since it requires that which is supposed be compatible with the context set). But it might happen that even when the speaker in fact knows that *A* is false, the addressee is not prepared to accept that *A* is false—it might be a point of contention in the conversation, in which case the assertion that the antecedent is false will not be accepted, and that proposition cannot be presupposed (by the speaker) without begging the question. In such a case, even though the speaker knows that the antecedent is false, she may have well-grounded conditional beliefs—well-grounded belief revision policies—which might constrain her conditional assertions.

For example, I take myself (correctly, let us assume) to know that Shakespeare wrote *Hamlet*. O'Leary, however, disputes it. It is common ground between O'Leary and me that the play was written by someone, and I take myself to know that it was, even conditional on the hypothesis that Shakespeare was not the author. Partly in the interest of ultimately convincing him that I am right about who wrote *Hamlet*, I might engage in debate with O'Leary about who the author of the play was or could have been if it wasn't Shakespeare—"Well, we can rule out Christopher Marlowe for the following reasons...". It makes sense to talk about conditional belief, conditional on hypotheses that are not only counterfactual, but contrary to the subject's knowledge. Can we also make sense of conditional *knowledge* in such cases? And if we can, should cases of conditional knowledge be reduced to cases of categorical knowledge of some kind of conditional proposition? Perhaps, but it is clear that it would be wrong to identify the relevant proposition with the corresponding counterfactual proposition. For while I take myself to know that even if Shakespeare didn't write *Hamlet*, *someone* did, I also take myself to know that if

explanation for this in terms of general conversational maxims. Similar explanations could be given by a defender of the conditional assertion account for why it would normally be inappropriate to make a conditional commitment on the basis of knowledge that the condition will remain unfulfilled.

Shakespeare hadn't written *Hamlet*, that play would never have been written.[18] We will return to the issue of conditional knowledge after putting our proposition-expressing hypothesis on the table.

(3) The commitment-to-defend hypothesis for the norm of categorical assertion suggests the following extension to a norm of conditional assertion: In the case of conditional assertions, one undertakes a commitment to defend the truth of the consequent against reasonable challenges, but only with one additional resource available to support it: the proposition expressed in the antecedent. This resource is not itself subject to challenge—the defense is conditional on it.

This account of a norm of conditional assertion also seems to yield the result that a conditional assertion is essentially the same, in the way it is assessed, as the categorical assertion of a material conditional. For if one could defend a material conditional against reasonable challenges, then one could use modus ponens to defend the consequent, with the additional help of the antecedent. And if one could defend the consequent with the help of the antecedent, then by a division of cases, one could defend the material conditional without the help of the antecedent.

(4) Finally, the Bayesian account of assertability suggests a natural extension to an account of conditional assertability, an account that has been one of the primary motivations for the development and defense of the non-propositional account of conditionals. If ordinary assertion goes by degree of belief, then conditional assertion goes by conditional degree of belief. Here, as proponents of the Bayesian account emphasize, we get a divergence between conditional assertability and the assertability of the material conditional, since the probability of a material conditional might be high (because of the low probability of the antecedent) when the conditional probability of consequent on antecedent is low (because the probability of the conjunction of the antecedent and consequent is much lower than the probability of the antecedent).

This completes our survey of the alternative accounts of norms of conditional assertion. We will look back at them after sketching a truth-conditional account that treats conditional propositions as propositions that are stronger than the material conditional.

3. Conditional Propositions

The propositional account I will sketch is the one given in my 1975 paper, "Indicative Conditionals."[19] The account is very simple, and has two components: first, a standard possible worlds and selection-function semantics for the conditional; second, a pragmatic constraint on the selection function that applies only to the case of indicative conditionals. The abstract semantics postulates a set of possible worlds and a selection function, f, taking a possible world α and a proposition A into a possible world, $f(A,\alpha)$. Propositions are represented by subsets of the set of possible worlds. The semantic rule for the conditional is this:

$(A > B)$ is true in possible world α if and only if B is true in $f(A, \alpha)$.

[18] Examples of this kind were first given by Ernest Adams. The Shakespeare example is from Jonathan Bennett.

[19] Reprinted as chapter 3 of Stalnaker 1999a.

The abstract semantics imposes a number of constraints on the selection function, motivated by the idea that the selection possible world should be a world in which the antecedent proposition is true, but that is otherwise minimally different (in relevant respects) from the base world from which it is selected. These constraints are common to conditionals of both kinds—indicative and subjunctive.

As we have noted, the pragmatic framework that provides the setting for an account of the effect of speech acts assumes that assertions occur in a context which determines a set of possible worlds—the *context set* that represents the live options among which speaker and addressee intend to distinguish between in their conversation. Propositions true in all of the worlds in the context set are (pragmatically) *presupposed* by the speaker. They include the information that is available, or that the speaker takes to be available, for the interpretation of utterances that occur in the discourse. If the language used is, in any way *context-dependent*, then the speaker will be assuming that the features of the context on which interpretation depends will be available, which requires that they be presupposed, which is to say that the information is entailed by the context set.

The second component of the account of indicative conditional propositions, a pragmatic constraint on selection function that determines the interpretation of conditionals, is as follows:

If A is compatible with the context set, C, then if $\alpha \in C$, $f(A, \alpha) \in C$.

The constraint is a partial specification of the respects of similarity and difference that are relevant to the interpretation of indicative conditionals: possible worlds compatible with the context are more similar to other possible worlds compatible with the context than they are to possible worlds outside the context set.

In the account of conditional assertion sketched above, we said that a supposition creates a *derived context*. There will also be a derived context on the propositional account (for both indicative and subjunctive conditionals). The derived context, $C(A)$ for conditional supposition, A made in context C, will be defined as follows: $C(A) = \{f(A, \alpha): \alpha \in C\}$. The effect of the pragmatic constraint is to ensure that in the case of indicative conditionals, the derived context set is a subset of the basic context set. Intuitively, this means that all the presuppositions of the basic context will be preserved in the derived context (assuming this is possible, which it will be provided that the supposition is compatible with the basic context set). The pragmatic constraint is motivated by the hypothesis that the role of the special morphology (the combination of tense, aspect and mood that distinguishes "If pigs could fly, they would have wings" from "If pigs can fly, they have wings," and that we have gotten into the habit of calling "subjunctive") is to signal that some of the presuppositions of the basic context are being suspended in the derived context. The idea is that the default assumption is that presuppositions of a basic context carry over to the derived context created by the supposition. In the absence of indication to the contrary, the pragmatic constraint will hold.[20]

[20] On this theory, both kinds of conditionals ("subjunctive" and "indicative") have the same abstract semantics, but a context-dependent parameter of the interpretation—the selection function—is differently constrained by the different grammatical constructions. So, on this theory, the difference between the two

It is required, on this account, that one use the so-called subjunctive morphology whenever presuppositions are being suspended, which is to say whenever the derived context is not a subset of the basic context. This implies that the conditional cannot be *counterfactual*, where what I mean by "counterfactual" is that the antecedent of the conditional is presupposed to be false. This does *not* imply, however, that speakers cannot suppose (in the indicative way) things that are incompatible with what they take themselves to know, or even to be common knowledge. All that is implied is that the conversational players must accommodate, expanding the basic context to include possible worlds compatible with the antecedent, if a speaker supposes something incompatible with the prior context. There is a difference between a counterfactual supposition (where we create a derived context disjoint from the basic context) and an "indicative" supposition of something incompatible with the prior context (where we expand the basic context so that the derived context can be a subset of it). To use the Shakespeare example again, when I say, "If Shakespeare hadn't written *Hamlet*, English literature would have been poorer than it actually is," I continue to presuppose, in the basic context, that Shakespeare did write *Hamlet*. But when I say, "If Shakespeare didn't write *Hamlet*, it must have been written by Marlowe", I adjust my basic presuppositions to accommodate the possibility that someone else wrote it. (I could not intelligibly say, for example, "If Shakespeare didn't write *Hamlet*, then it was written by a different person than the one who actually wrote it.")

The combination of the selection function semantics with the pragmatic constraint on selection functions yields an overall account of indicative conditionals that reconciles the thesis that the proposition expressed by a conditional is stronger than the material conditional with the fact that the acceptance of the material conditional seems to be sufficient for the acceptance of the corresponding indicative conditional (at least in contexts in which the negation of the antecedent was not accepted). Once a disjunction of the form (*not-A or B*) is accepted in a context (becomes part of the common ground), then the pragmatic constraint on the selection function ensures that the corresponding indicative conditional (*if A, then B*), will be true in all of the possible worlds in the context set, and so it will be accepted as part of the common ground that the indicative conditional is true. Dorothy Edgington, in an influential general critique of truth-conditional analyses of conditionals, used the prima facie conflict between this thesis about conditional propositions (that they entail, but are not entailed by, the corresponding material conditional) and this fact about the acceptance conditions for indicative conditionals (that acceptance of the material conditional is sufficient for the acceptance of the indicative conditional) to argue against the thesis. She acknowledged that the account I had developed provided a way around her criticism, but she argued that it

kinds of conditionals is a *semantic* difference in two different senses, but a purely *pragmatic* difference in a third sense. The difference is semantic, first in the sense that there will normally be a difference in the proposition expressed by the contrasting conditional sentences, even when uttered in similar situations. And it is semantic also in the sense that the difference is marked by a conventional linguistic device (the tense/aspect/mood difference). But the distinction is pragmatic in that the device works by the way it constrains features of the context. The semantic rule that gives the truth conditions of the conditional as a function of the contextual parameter will be the same for both kinds of conditionals.

did so at too high a cost, the cost of making indicative conditionals unacceptably context-sensitive. Her criticisms of my account do point to real problems, but in detail, they are off the mark, missing two distinctions that are important, whatever the ultimate fate of the truth-conditional analysis.

Edgington claims that this analysis gets *acceptance* conditions right only "by making 'truth' and 'truth conditions' radically information-dependent." Specifically, she claims, "If one party is certain that if *A*, *B*, and another is not (but regards *A* as possible), they *cannot* be disagreeing about the obtaining of the same truth conditions. They *must* be equivocating. For the former's context set must rule out *A & ~B*, and the latter's must not." But this is not correct: it confuses the context set (which, in a non-defective context is the same for speaker and addressee) with the beliefs of the speaker and addressee (which will always be different, if they have any reason to communicate with each other). The pragmatic constraint applies only to the context set—the common ground, which in a non-defective case, coincides with the possible situations compatible with what is presupposed. Even if one party is certain that if *A*, then *B*, if the other is not, and if it is recognized by both that at least one party to the conversation is not in a position to exclude the possibility that *(A&~B)*, then it will not be *presupposed* by either party that if *A*, then *B* (or that *~(A & ~B)*). The two may still understand the conditional in the same way.

Edgington may reply that one cannot fully reconcile the truth-conditional account with the phenomena about the acceptance conditions for indicative conditionals unless one extends the pragmatic constraint so that it applies to the private beliefs of speakers as well as to the common ground. The fact she appeals to is that all one needs to know in order to be in a position to make a conditional assertion (in a situation in which one is uncertain whether the antecedent is true) is that the material conditional is true. To account for this fact, on the assumption that the conditional expresses a proposition with a selection-function semantics, one must assume that the speaker's selection function will give priority to possible situations compatible with her beliefs, as well as to possible situations compatible with the common ground. But we can accept this extension of the constraint while still denying that speakers with relevantly different beliefs are equivocating or misunderstanding each other. What must be granted is that in some cases, indicative conditionals are implicitly about the speaker's beliefs. We must allow that what I say when I say something of the form "*if A, then B*" may not be the same as what you would have said, using the same words. But whether one is defending a truth-conditional account or a conditional assertion account, one needs to recognize that indicative conditionals are conveying information about the speaker's epistemic situation. On either kind of account, we should distinguish *assertion* conditions (conditions under which one is in an epistemic position to make a conditional assertion) from *acceptance* conditions (conditions under which one is in a position to accept a conditional assertion made by someone else.) I may be in a position to *assert* a conditional while at the same time be prepared to *accept* one that appears to conflict with it, if asserted by you.[21]

[21] Here I am indebted to the comments of the editors of *New Work on Modality* (where an earlier version of this chapter was published) for helping me to see the issue more clearly.

Once we acknowledge that our indicative conditional statements are in part about our individual epistemic situations, then one thing that needs to be explained is the fact that conditional statements with the same antecedent and contrary consequents that are made by different speakers in the same context seem to conflict with each other. I think that our account of proposition-expressing indicative conditional statements, together with an account of the dynamics of conversation, can account for such conflicts, as well as for cases where one first *accepts* one conditional, and then *asserts* one that seems to conflict with it.

What I *would have said* had I said "*if A, then B*" may be different from what you say with the same words, but once you have said it, the context changes, and if I still am unsure about the truth of the antecedent, I cannot say "*if A, then not-B*" without disagreeing with you. On the other hand, if I learn from what you say that the antecedent is false, then I may accept the truth of your conditional, while continuing to affirm one that appears to conflict with it. The following example illustrates this phenomenon. You say, "One if by land, two if by sea," meaning that if the British are coming by land, there will be one lantern in the tower, and if they are coming by sea, there will be two lanterns in the tower. I *accept* what you say, and knowing already that there is just one lantern in the tower, I conclude that the British are coming by land. Before, I was in a position to *assert,* "If the British are coming by sea, there is still only one lamp in the tower," and I am still prepared to say this. If the British are coming by sea, then there must have been some mistake about the signal. There is context-shifting and radical information-dependence here, but a truth-conditional account of what is going on can explain the phenomena, and does not have to say that there is equivocation or misunderstanding.[22]

Edgington's second complaint is that the pragmatic constraint requires an anti-realist conception of truth:

> If a context set is sufficiently bigoted or bizarre, any old (non-contradictory) conditional can come out "true": "If we dance, it will rain tomorrow," for instance. We dance, and the drought continues unabated. Given what we now know, we would not have uttered those words. But the context is different—in its own context, what was said was "true."

But this complaint is misguided, equivocating on what is meant by saying that something is true in a context. A context, in the general framework in question, is represented by a set of possible situations—the context set—which encodes the information that speaker and addressees take to be common ground. Context-sensitive utterances will have truth conditions that are sensitive to the context set, so that there are utterance that might have had different truth values (in the actual world) if their contexts had been different. For example, in one context it is common ground that the speaker intended to refer, with "that man" to George W. Bush when she said, "That man won the election," while in another context, the speaker uttered the same sentence, but there it is common ground that he intended to refer to John Kerry. One might say that the sentence was true in the first context, and false in the second. This is the most straightforward sense of "true in a context". On the other

[22] The Paul Revere example is discussed briefly in Stalnaker 1984, 108.

hand, since the context is represented by a set of possible worlds, one might mean by "true in a context" what is true in all of the possible worlds in the context set that represents that context. On this interpretation, to say that something is true in a context is to say that it is presupposed in that context—which is to say that it was taken to be common ground (whether it is actually true or not). In the second sense, "That man won the election" will be true in both of the above contexts, assuming that the assertion was accepted in both contexts. Truth in a context, in the second sense, is not a special notion of truth, but is just truth with respect to certain nonactual situations.

Edgington's rain dance example is an example of a conditional that is presupposed, and presumably believed, to be true, but that is in fact false. So, it is "true in the context" only in the second (somewhat misleading) sense. This is no more a problem for a realist notion of truth than the fact that "John Kerry won the election" is true in possible situations compatible with the beliefs of people who think that he won. The pragmatic constraint on selection functions does imply that the actual truth of a conditional will sometimes be context-sensitive, but only when the actual world is compatible with the context, since it constrains the truth-conditions of the conditionals only in the possible worlds in the context set—those compatible with what is presupposed.[23] The constraint has no consequences for the truth-conditions of conditionals in possible worlds outside the context set, including the actual world, should that world be incompatible with what is presupposed.[24] But while the pragmatic constraint is not relevant to the actual truth value of the conditional in Edgington's example, the abstract semantics requires that it be false with respect to any context, since the antecedent is true, and the consequent false.

I have emphasized, in this response to Edgington's objection, that the pragmatic constraint that I proposed has consequences only for the truth conditions of conditionals in possible worlds that are compatible with what is presupposed in the context (possible worlds in the context set). One might worry that this constraint is therefore much too weak, saying nothing, in most cases, about the actual truth of indicative conditionals. The worry is that it seems reasonable to believe that in almost all contexts, speakers will be making at least one false presupposition, even if an irrelevant one. Consider a discussion between two creationists about the Kennedy assassination. They falsely presuppose the truth of some creationist doctrines, let us assume. Even though these presuppositions are irrelevant to their current discussion, they imply that the actual world will be outside of the context set, so the constraint will be silent on the *actual* truth value of a statement like "If Oswald didn't shoot Kennedy, someone else did". But it seems clear that this statement, made in such a context, would be true, and the explanation for why it is true should be the same as in the case where all the presuppositions are true.[25]

[23] The truth value of a conditional statement will presumably be context-sensitive in some cases, even when the actual world is outside the context set, but my point is that this one specific pragmatic constraint does not imply that it is.

[24] This reply to Edgington criticism is developed in Block 2008.

[25] This concern was pressed by the editors in their comments. The creationist example is theirs.

There is a more general issue here concerning the assessment of the truth-value of statements, particularly context-sensitive statements, that are made in a context in which some presuppositions are false.[26] One way to limit the effect of irrelevant presuppositions (a way that is independently motivated) is to adopt a more coarse-grained analysis of the possible situations that define a context set.[27] The possibilities, on this kind of account, will be cells of a partition of the space. Possible worlds will be members of the same partition cell if they are equivalent in all respects that are relevant to what is at issue in the context. The actual situation will be the partition cell that contains the actual world. In a model of this kind for the creationist example, the actual situation will be compatible with the context, since the presuppositions that are relevant to the current discussion are all true.

So, this truth-conditional account of indicative conditionals is not quite so quickly refuted as Edgington thought, and does not require any kind of nonstandard conception of truth. No truth-conditional account will have a problem with examples (such as Edgington's rain dance example) with true antecedents. But there are related problems, brought most clearly into focus by Allan Gibbard's famous example of Sly Pete, the Mississippi Riverboat gambler. This is an example of a pair of apparently contrary conditionals with false antecedents, but which are made in a context in which the actual situation is compatible with what is presupposed. Here is Gibbard's story:

> Sly Pete and Mr. Stone are playing poker on a Mississippi riverboat. It is now up to Pete to call or fold. My henchman Zack sees Stone's hand, which is quite good, and signals its contents to Pete. My henchman Jack sees both hands and sees that Pete's hand is rather low, so that Stone's is the winning hand. At this point the room is cleared. A few minutes later Zack slips me a note which says 'if Pete called, he won,' and Jack slips me a note which says 'if Pete called, he lost.' . . . I conclude that Pete folded.[28]

Jack and Zack, in this story, each have only partial information about the situation but each recognizes the way in which his information is partial, and neither is making any mistake. So, the actual situation is compatible with the beliefs of both henchmen, and we may presume it is also compatible with the contexts in which each communicates with the narrator of the story (Call him "Allan"). So, the putative problem for the truth-conditional account is this: Suppose that indicative conditionals express propositions, and that a conditional assertion is the categorical assertion of a conditional proposition. Then it seems that (1) Zack is not in a position to rule out what is in fact the actual situation in which Pete folds with a losing hand, and it does not seem that he is ruling it out. But (2) it seems that he is in a position to assert "if Pete called, he won" (thus ruling out all possible situations incompatible with the truth of the proposition expressed). But then it must be that (3) the conditional

[26] See Yablo 2006 for a take on the general problem.

[27] I am drawing here on the work of Seth Yalcin on what he calls "modal resolution". See Yalcin 2008. See also work by Jonathan Schaffer (for example, Schaffer 2007) who argues that we need to enrich the idea of a context to include a partition of the space of possibilities that is determined by the alternative answers to the questions that are at issue in the context.

[28] Gibbard 1981, 231.

proposition expressed is true in the actual situation. Similarly for Jack, with his assertion of an apparently contrary conditional proposition. So, it seems that when Jack says, "If Pete called, he lost," he speaks the truth, but also that Zack speaks the truth when he says, "if Pete called, he won". If we further assume that the conditional sentences are interpreted with a selection-function semantics, then we must assume that the selection function relevant to interpreting Zack's conditional is different from the one relevant to the interpretation of Jack's. To explain the truth and assertability of the conditionals, each selection function must be constrained, not only by the presumed knowledge that they share—the common ground—but also by their private beliefs or knowledge. But the information necessary to interpret a context-sensitive expression is supposed to be presumed by the speaker to be available to the addressee, and so to be presupposed. In general, a speaker's private beliefs and knowledge may be expressed in his assertions, but where they are not common ground, they cannot be presumed to be available for the interpretation of what is expressed.

But as we have seen, there is no conflict between the assumption that the selection function for interpreting conditional propositions is constrained by a speaker's private beliefs and the assumption that the relevant selection function must be public knowledge. As I suggested above, in discussing Edgington's first objection, the assumption that the selection function is constrained by the speaker's private beliefs *does* imply that a conditional in one speaker's mouth may say something different from what would have been said by a different speaker, in the same context, with the same conditional sentence. But it does *not* imply that the information necessary to interpret each speaker's conditional is not publicly available. For if it is a general constraint that selection functions for indicative conditionals should, where possible, be closed under the speaker's knowledge or beliefs, then it will be common ground that this constraint will hold. Even if the addressee does not know what the speaker knows, he will know, and it will be common ground, that the world selected from a given possible world in the context set will be constrained by what the speaker knows or believes in that possible world.

Let me spell the point out in detail, in terms of the riverboat example[29]: Let α, β and γ be three possible worlds in which Pete in fact folded. α is the actual world in which Jack knows that Pete had a losing hand. β is a possible world in which Jack knows that Pete had a winning hand, and γ is a possible world in which Jack does not know whether Pete has a winning or a losing hand. Let f_J be the selection function relevant to interpreting Jack's conditional statements, let *P* be the proposition that Pete called, and let *L* be the proposition that he lost. Possible worlds of the three kinds (α, β and γ) are compatible with the knowledge of Allan, the narrator, and with the context in which Jack's message is passed to Allan. But even though Allan does not know, and Jack does not presuppose, that the actual world is α, rather than β or γ, Allen does

[29] In Gibbard's telling of the story, he adds that Allan knows that the notes come from his trusted henchmen, but doesn't know which note came from which (and so presumably does not know anything about the basis for their conditional claims.) To simplify the analysis (avoiding a proliferation of alternatives), I assume that Allan knows that the note came from Jack, and something about Jack's epistemic situation. I don't think anything essential is lost by this simplification of the story.

know, and it is common ground, that $f_J(P,\alpha)$ is a possible world in which L is true, and that $f_J(P,\beta)$ is a possible world in which L is false. (Nothing is implied by the constraints we are considering about $f_J(P,\gamma)$.) When Allan receives Jack's message, he understands it, and when he accepts it, he rules out possible world β. (He also, of course, rules out possible worlds in which Pete called and won.)

What about possible worlds like γ? Suppose that, in γ, Jack does not know anything that gives him reason to exclude either the possibility that Pete called and won, or the possibility that Pete called and lost. Then nothing we have said implies that the literal content of Jack's conditional statement is true, or that it is false, but Allan will still be able to exclude that possibility on the ground that Jack would not be in a position to assert the conditional if world γ were actual.

Even though Jack would not have made the conditional statement if world γ were actual, we still might ask whether (on the truth-conditional account) the conditional proposition he expressed (in α) was true or false in γ. It seems intuitively clear that the corresponding *counterfactual* conditional proposition, "if Pete had called, he would have lost" will be true (whatever Jack knew or believed) just in case Pete had a losing hand, but what about the indicative conditional? Here, the most natural intuitive judgment may be that the question does not arise—no non-arbitrary truth value can be assigned. This may seem to lend support to the conditional assertion account, according to which nothing has been asserted when the antecedent is false, but a proponent of the truth-conditional account might say the same thing, allowing (as seems plausible for both counterfactual and indicative conditionals) that selection functions may sometimes be only partially defined. A cautious version of the kind of truth-conditional account of indicative conditionals that I have been promoting might hold that, with indicative conditionals, *all* selection functions compatible with the epistemic and contextual constraints are admissible, and that indicative conditionals are true if and only if true with respect to all admissible selection functions, and false if and only if false for all. This version of the truth-conditional account will be essentially equivalent to a version of the conditional assertion account.

It is not that I want to defend this cautious version of the truth-conditional analysis, which implies that non-trivial indicative conditionals will be true or false *only* in possible worlds that are compatible with what is presupposed in the context. The point is just to note that we can see the conditional assertion account as equivalent to a limiting case of the truth conditional account. One may not have to choose between the two alternatives: it may be that one can have some of the advantages of both. One advantage, in particular, to bringing the truth-conditional and conditional assertion accounts together is that it may facilitate a more unified theory of conditionals, and an explanation of the relations between indicative and "subjunctive" conditionals.

To account for the phenomena, we must assume, as Edgington noted, that conditionals are information-sensitive, both to public and private information. The pragmatic constraint imposed in my original account of indicative conditionals concerned only the relation between common ground—publicly available information—and the interpretation of conditionals. The aim there was limited to an explanation of the force of a certain argument, and more generally, to an exploration of one case of the interaction of context and content in the dynamics of discourse. But the

Gibbard example, and others, make clear that indicative conditionals, if they are given a truth-conditional account, will be sensitive also to the knowledge or beliefs of individual speakers and thinkers, and so to information that is not, in general, publicly available. This kind of speaker-relativity means that indicative conditionals are, implicitly, in part about the speaker's epistemic situation, but this does not require any non-standard notion of proposition, or of truth, and it does not imply that speakers and addressees who are ignorant of each other's private beliefs do not understand each other.

The two kinds of constraints on the interpretation of indicative conditionals (by the common ground, and by the speaker's individual epistemic situation) will interact, and there will in some cases be tensions between them. I will conclude by considering an example that illustrates one such tension—an example that raises a problem for both truth-conditional and conditional assertion accounts. I am not sure what to say about this example, but will make a tentative suggestion.

4. Public vs Private Information

There are only three possible suspects for the murder—the butler, the gardener, and the chauffeur—and it is common knowledge that whoever did it acted alone. Alice was with the gardener at the time of the murder, so she is absolutely certain that he is innocent. Bert has conclusive evidence that rules out the chauffeur, which he has shared with Alice, so it is common knowledge between Alice and Bert that either the butler or the gardener did it. Alice concludes (privately) that since it wasn't the gardener, it must have been the butler. But Bert has what is in fact misleading evidence that he takes to exonerate the butler, so he infers that it must have been the gardener. Alice and Bert each tell the other who he or she believes was the guilty party, but neither is convinced by the other. Alice, in particular, is far more certain of the innocence of the gardener than she is of the guilt of the butler; were she to learn, to her surprise, that the butler was innocent, she would conclude that the chauffeur's alibi must not be as good as it looks, and that he is the guilty party. But that won't happen, since in fact, the butler did it.

Bert says, "We disagree about who did it, but we agree—it is common knowledge between us—that either the butler or the gardener did it, and each possibility is compatible with our common knowledge. So even though you are convinced that the butler is the guilty party, you should agree that if the butler didn't do it, the gardener did." Bert is just giving what I have called the direct argument, which all of the accounts on the table have assumed to be compelling.

Alice agrees that it is common knowledge that either the butler or the gardener did it, and that each of the two possibilities is compatible with their common knowledge. But she will be reluctant to accept the conditional, which conflicts with her conditional belief—perhaps with her conditional knowledge—that even if the butler didn't do it, the guilty party is still not the gardener.

If, in this case, we assume that common knowledge (what Alice and Bob both know, know that they know, know that they know that they know, etc.) coincides with the common ground (the context set), then if Alice reasonably refuses to accept Bert's conditional conclusion ("if the butler didn't do it, the gardener did"), we have a counterexample to the pragmatic constraint imposed in my original truth-conditional account of indicative conditionals. The example also seems to conflict with our conditional assertion account, which implies that one should accept *B*, conditional on *A*, if *A*, added to the common ground, entails *B*. But on the other hand, if Alice accepts the conditional that Bert invites her to accept, she will be apparently violating a conditional knowledge norm, and (from the point of view of the Bayesian account of assertability) asserting something unassertable.[30] On the truth-conditional analysis, what we seem to have is a case where the requirement to select a possible world from the context set conflicts with the requirement to select a possible world from the set of possibilities compatible with the individual's conditional knowledge.

I am not sure what the best response to this problem is, but I am inclined to think that one should question the assumption that everything that is common knowledge is common ground. That is, I am inclined to think that some possible situations that are incompatible with the common knowledge of the parties to a conversation are nevertheless "live options" in that conversation. When it becomes clear that the guilt of the butler is in dispute, so that it is a live option in the context that the butler didn't do it, Alice should insist that we reopen the possibility that the guilty party was the chauffeur. ("I am sure it was the butler, but if it wasn't the butler, it might have been the chauffeur.") The context set should be expanded to include possibilities compatible with the conditional knowledge of the parties on any condition compatible with the context. This would be a way of reconciling the conflicting constraints, and seems to be intuitively plausible.[31]

The riverboat example was an example of the pooling of information. The different parties knew different things, but had no disagreements. The murder example is a case where the different parties know what each other thinks, but they disagree, so their beliefs continue to diverge. This kind of case may require a more complex account of the common ground.[32]

[30] I assume that in an appropriate Bayesian account, conditional probabilities may be defined even when the probability of the condition is 0. Such conditional probabilities represent conditional degrees of belief, on conditions that are entertainable, even though they are taken to be certainly false. Alice is disposed, should she be surprised by the information that the Butler was certainly innocent, to revise her beliefs so that she would still assign very low credence, relative to that condition, to the proposition that the gardener did it.

[31] Thony Gillies pointed out that, if we expand the context set in this way, then Bert could attempt to close it again simply by asserting the disjunction, "either the butler or the gardener did it." Alice, it might seem, should accept this disjunctive assertion, since its content is entailed by something she accepts. But the response I am tentatively proposing must say that Alice should reject the assertion, despite the fact that it is entailed by something she believes. Given that her assertion that the butler did it was rejected, she should not be willing to accept this particular weakening of her assertion. "If we can't agree that the butler did it, we also can't agree that it was either the butler or the gardener." It is surprising that one might reasonably reject an assertion even if one would assert or accept one that entails it, but this still may be the right response. Compare Grice's discussion of what he calls "substitutive disagreement", where one rejects a disjunctive statement, "either Wilson or Heath will be the next Prime Minister" in favor of another that shares one of the disjunctions ("I disagree. It will be either Wilson or Thorpe.") (Grice 1989, 64).

[32] Thanks to Thony Gillies, Isadora Stojanovic, Andy Egan, and the editors of *New Work on Modality* (where this chapter was originally published) for helpful comments on earlier versions of this paper. Thanks also to the editors of this volume, Andy Egan and Brian Weatherson, for their helpful and stimulating comments and suggestions. Finally, thanks to Jonathan Bennett for conversation and correspondence over the years about conditionals that helped me to get clearer about many issues.

11

Counterfactuals and Probability

1. Introduction

Dorothy Edgington has long defended what Jonathan Bennett labeled the NTV thesis[1] about conditionals: conditional sentences do not express propositions, but should be understood as expressions of conditional belief, where a conditional belief is qualified belief in the consequent proposition, with the antecedent proposition determining the qualification. Edgington developed her account of conditionals in a probabilistic context: conditional belief is conditional credence, or degree of belief, modeled by conditional probability. Her NTV thesis contrasts with the thesis, defended by David Lewis, among many others, that a conditional sentence expresses a proposition that is a function of the propositions expressed in the antecedent and consequent clauses (plus context).

Edgington has also long argued that the contrasting kinds of conditionals—standardly but inaccurately labeled subjunctive and indicative[2]—should receive a unified treatment. While all acknowledge that there are contrasts between indicative and corresponding subjunctive conditionals, including cases where the indicative is assertable, while the corresponding subjunctive is not (and vice versa), Edgington argues that in many cases, the subjunctive can be interpreted as a restatement, in a retrospective context, of what was said with an earlier indicative. More generally, she argues that we should explain the differences between the two kinds of conditionals within a common framework. On this issue, her account of conditionals contrasts both with some propositionalists, such as David Lewis and Frank Jackson, who argue that indicative and subjunctive conditionals should receive fundamentally different propositional analyses, and also with theorists such as Allan

[1] In Bennett 2003, Chapter 7.

[2] The "subjunctive" label is grammatically inaccurate since while some so-called subjunctive conditionals in English do involve the grammatical subjunctive, it is tense and aspect rather than grammatical mood that are doing the work of marking the distinction. The label "indicative" is also inaccurate, not only because some so-called subjunctive conditionals are literally indicative, but also because at least in archaic English, some conditionals that are literally (present tense) subjunctive should be grouped, by meaning, with the so-called indicative conditionals. (For example, "If he be found guilty, he will appeal.") It is hard to find labels for the distinction that are both grammatically accurate and theoretically neutral ("counterfactual" is problematic for a different reason), but there does seem to be a robust distinction clearly marked in the grammar, and general agreement about paradigm cases of the contrast. I will follow tradition, as Edgington does, in using the subjunctive/indicative terminology, but these terms should be understood to mean "so-called subjunctive" and "so-called indicative". See the chapter by Sabine Iatridou in this volume for an interesting discussion of grammatical issues involving conditionals, as well as an earlier paper, Iatridou 2000.

Gibbard, who defend the NTV thesis for indicative conditionals, but a propositional analysis for subjunctive conditionals.[3]

On the first issue, I have been on the propositionalist side, while acknowledging that there is some, perhaps considerable, indeterminacy in the application of the truth-conditional semantics. That is, it is acknowledged that conditional sentences may often express only partial propositions. (If we identify a *proposition* with a function from possible worlds to truth-values, then a partial proposition is one that is defined for some but not all possible worlds.) On the second issue, I have been with Edgington on the unificationist side, aiming to explain the differences between the two kinds of conditionals as variations of a single semantic analysis. The two issues interact, since the NTV thesis is prima facie more plausible for indicative conditionals, while the propositionalist's case is strongest for subjunctive conditionals. So, one may use the unification hypothesis, as Edgington does, to generalize the NTV thesis from indicative to subjunctive, or as I have, to motivate a propositionalist account of indicatives.

While I have been on the propositionalist side on the first issue, I have also argued that the distance between a conditional assertion account of indicative conditionals of the kind that Edgington has developed and the propositionalist account of indicative conditionals that I have defended is less than might appear. Specifically, I have argued that a version of the latter is essentially equivalent to a version of the former. One can see the conditional assertion account as a limiting case of a propositionalist account, and this helps to facilitate an explanation of the continuity between the two kinds of conditionals, and so to support a unified account. One of the things I want to do in this chapter is to develop further the ecumenical attempt at reconciliation, this time focusing on the contrasting accounts of subjunctive conditionals. I also will try to get clearer on exactly what is at stake in the debates about whether conditionals express propositions, and whether they have truth-values.

Here is my plan: In section 2, I will review my attempt to reconcile a conditional assertion account of indicative conditional speech acts with a truth-conditional analysis of conditional sentences, connecting it with a more general project of reconciling a truth-conditional semantics for deontic and epistemic modals such as "may," "might," and "must" with an expressivist account of the role of modal statements in discourse. But even if one *can* give the conditional assertion account of indicative conditionals a truth-conditional formulation, one may ask why one *should*—what advantages this kind of formulation offers. So, in section 3, I will try to motivate the truth-conditional formulation by appealing to continuities between the conditionals that best fit the NTV account and those that seem to be used to make factual claims. Edgington herself appeals to these continuities, using them to argue for extending the NTV account of indicative conditionals to counterfactuals, but I will argue in section 4 that her way of connecting indicative conditionals with counterfactual statements that are made in a later retrospective context is in tension with the thesis that indicative conditionals express conditional probabilities. Her defense of an NTV account of counterfactuals also appeals to the intuitive judgment

[3] Gibbard 1981.

that some counterfactuals should get probability values, but not truth-values. I agree with her intuitive judgments about the cases in question, and will try, in section 5, to reconcile these judgments with a broadly truth-conditional analysis of counterfactual conditionals.

2. Expressivism about Indicative Conditionals and Epistemic Modals

The truth-conditional semantics that I proposed for conditionals in general posited a selection function that takes two arguments (a possible world and a proposition) to a possible world.[4] The intuitive idea was that if w is the actual world, then $f(w,\phi)$ is a world that differs from w only in the minimal way required to ensure that ϕ is true. Structural constraints were imposed on the function to ensure that it conforms to this idea, but no attempt was made, in the abstract semantics, to give any substantive criteria for determining what counts as a minimal difference, or a "closest" possible world. That, it was assumed, depended on the application of the semantics, and on substantive metaphysical and epistemological hypotheses. My aim in proposing this semantic analysis was more modest than the aim of David Lewis, who independently proposed a formally similar semantic analysis at about the same time.[5] Lewis intended his analysis (which was taken to apply only to the so-called subjunctive conditionals) to be part of a reductive project in the spirit of Hume, who Lewis described as the great "denier of necessary connections." I am skeptical of Lewis's reductive project, and of the grand metaphysical thesis that he labeled "Humean supervenience," but that is an issue that can be separated from the question whether a truth-conditional analysis is appropriate for conditionals.[6]

In contrast with Lewis, I took this kind of abstract truth-conditional analysis to apply to both indicative and subjunctive conditionals, but it was assumed that the grammatical marks that distinguish the two kinds of conditionals signaled a difference in the criteria by which minimally different possible worlds are selected. In the case of the indicatives, the criteria are epistemic, and I proposed the following constraint on the selection function: for possible worlds compatible with what is presupposed, the selected worlds must also be compatible with what is presupposed. That is, if C is the *context set*—the set of possible worlds compatible with what is presupposed in the relevant context, then a selection function f is *admissible* for the interpretation of an indicative conditional if and only if for any $w \in C$, $f(w, \phi) \in C$. (Since it is a general constraint on selection functions that $f(w, \phi) \in \phi$, the definition of an admissible selection function presupposes that indicative conditionals are interpreted only when the antecedent is compatible with the context set.) There will be many admissible selection functions. The constraint is just that the function or functions relevant to the interpretation of indicative conditionals must be admissible.[7]

[4] Stalnaker 1968 and Stalnaker 1984, chapter 7.

[5] Lewis 1973.

[6] I criticized Lewis's reductive project in Stalnaker 1984, chapter 8, and more recently in more detail in Stalnaker 2015.

[7] Stalnaker 1999a, ch. 3 (originally published in 1975).

I say function *or functions* because it was recognized from the start that in application, the context in which a conditional of any kind is interpreted may not fully determine the parameter relative to which the formal semantics specifies truth-conditions for the conditional. The idealized semantics makes a uniqueness assumption: for each proposition, there is a unique possible world (or possible situation) that is *the* possible situation in which the proposition is true, and that is minimally different from the actual situation (or more generally, the situation relative to which the conditional is being evaluated). But in practice, the relevant context may provide only constraints on the parameter that do not fully determine it. The gap between ideal semantic theory and practice is not a special feature of conditionals: any contextual parameter (a domain of discourse for the interpretation of a quantifier, a reference class for the interpretation of a gradable adjective) might be only partially determined. A supervaluation strategy[8] is used to model this kind of gap between theory and practice: a statement is true if true for all precisifications of the partially determined parameter, false if false for all, and neither true nor false if true for some and false for others. The supervaluation strategy allows compositional semantics to proceed at the level of ideal theory, with supervaluations determined for the result.[9]

If we have models that allow for a class of selection functions, rather than a single selection function, then we can consider the following limiting case of my account of indicative conditionals: let the class of selection functions be the class of *all* selection functions that are admissible, relative to the given context set. In a model of this kind, a conditional will be true in a possible world compatible with the context if and only if the consequent is true in all possible worlds in that context in which the antecedent is true, false if false in all of those possible worlds, and neither true nor false otherwise.

This limiting case is the maximally cautious, or minimally committal interpretation that fits the truth-conditional analysis of indicative conditionals sketched

[8] See van Fraassen 1966, where the supervaluation idea is first introduced.

[9] Supervaluations are just one part of a multi-pronged strategy for reconciling the uniqueness assumption in the formal semantics with the manifest fact that it is hopelessly implausible to assume that we have what the ideal theory seems to require: a well-ordering of all possible worlds relative to each possible world. The implausibility is lessened when one recognizes that the semantics need not assume that the entities labeled "possible worlds" are metaphysically complete ways a world might be, but can be relevant alternative possibilities, individuated by what is at issue in a particular application. One might think of the possibilities as partition cells of a partition of logical space, rather than point in the space, and where the partition is relatively coarse-grained, it not so implausible to suppose that we have a function that selects a unique one that is minimally different from the actual situation. But it should also be said that we need not assume that even in such a limited context, conditionals are defined for all possible antecedent suppositions. One may think of the antecedent of a conditional as something like a definite description. The speech act of supposing that ϕ might be thought of as something like a request to consider *the* minimally different possible situation in which ϕ. The supposition is appropriate (just as a singular definite description is appropriate) only in contexts that provide a unique salient situation, or referent.

The need to jump through various different hoops to save the semantics that validates the principle of excluded middle raises a methodological concern: are these maneuvers just ad hoc devices to save a theory? But whether one should worry about that depends partly on whether the linguistic evidence supports the validity of the principle of excluded middle, as I think it does, and also on whether the necessary maneuvers have independent interest and motivation, which I think they do.

above. If we combine this truth-conditional semantics with an account of speech act force, we get a general account that is essentially equivalent to a conditional assertion account. To make an assertion, on the general account of assertion that I have given, is to propose that the propositional content of the sentence asserted (in the relevant context) be added to the common ground. With ordinary assertions that purport to convey factual information, the sentence is interpreted in the prior context, and the proposition determined relative to that context then is added to determine a posterior context (assuming the assertion is accepted). But in some cases, including deontic and epistemic modals, one may interpret the sentence in a prospective way, relative to the *posterior* context. The speech act rule will be something like this: adjust the context in the minimal way required to make the sentence true relative to the resulting context. The truth-conditional semantics plays an essential role in the account of speech acts of this kind, but the result helps to explain their expressive or practical role: the sense in which a statement does not make a factual claim, but expresses an attitude other than categorical belief. So, for example, one can take a statement like "you may have a piece of pie," not as stating the prior fact that having a piece of pie is permissible, but as a speech act of giving permission. One can understand "the butler might have done it" as an expression of the fact that the speaker is not in a position to rule out the possibility that the butler did it. The "might" statement, on this account, is an expression of the fact that the speaker is not in a position to rule out the possibility, but it is not a statement of that fact. It can be understood as a proposal that the possibility not be excluded from the context. If we take indicative conditional assertions to be speech acts of this kind, with the truth-conditional semantics I have sketched, then we have an account that can be seen as a formulation of a conditional assertion analysis: an account that says that one is in a position to assert (if ϕ, ψ) if and only if one is in a position to assert ψ in a derived context obtained by adding the condition ϕ to the prior context. In making a conditional assertion, one is expressing a conditional belief.[10]

Of course, beliefs and conditional beliefs can be partial, as Edgington's work emphasizes. Her account began with Adams's thesis: that a conditional is assertable if and only if the conditional credence of consequent on antecedent is sufficiently high. I will look at some putative counterexamples to Adams's thesis in section 3, but for now I want just to make two points about the relation between Adams's thesis and our truth-conditional formulation of a conditional assertion account. First, our account of conditional assertion does not *entail* Adams's thesis: the conditional assertion account says that one is in a position to assert "if ϕ, then ψ" in a context C if and only if one is in a position to assert ψ in a derived context defined by adding (temporarily) the supposition ϕ to the context C, but nothing is said about the conditions in which one is in a position to make an assertion in a given context, either a basic context or a derived one that is created by a supposition. The hypothesis that conditional assertion goes by conditional probability is an additional element of the theory, one that adds a Bayesian norm of assertion.

[10] See Stalnaker 2014, chapter 6, for discussion of this kind of account of deontic and epistemic modals, and indicative conditionals. The basic idea starts with David Lewis's game of commands and permissions. See Lewis 1975.

The second point is that the truth-conditional formulation of the conditional assertion account of indicative conditional statements is *compatible* with Adams's thesis. That is, the truth-conditional analysis is compatible with a further hypothesis that one is in a position to make a conditional assertion if and only if one's conditional credence of consequent on antecedent is sufficiently high. One might be tempted to think that David Lewis's notorious triviality proofs[11] showed that one cannot consistently combine Adams's thesis with a propositional analysis of the conditional, but this is not correct. Lewis's proof (and most of the many variations of it) was a reductio with two assumptions: (1) that conditionals express propositions, and (2) that Adams's thesis holds for a nontrivial class of probability functions that is closed under conditionalization. (A class of probability functions is defined as nontrivial if it includes at least one probability function that assigns non-zero probability to at least three disjoint propositions.) But there was also an implicit assumption, which Bas van Fraassen identified, and labeled "metaphysical realism."[12] The implicit assumption is an invariance condition: roughly, that the proposition expressed by a conditional sentence does not vary with a change in the context, where a credence function is taken to be an essential part of the context. Conditionalization is a shift from one credence function to another, and so from one context to another. The shift is from a prior belief state to the posterior state one would move to upon receiving the information that is the content of the condition. Suppose we label the prior credence function "P," and the posterior credence function that would result from learning B and conditionalizing "P_B." Assume that the arguments of the probability functions are *propositions*—the ones expressed by the sentences, A, B, A→B, etc. Then for any proposition X, $P_B(X) =_{df} P(X/B)$. A crucial step in Lewis argument assumes that $P_B(A \rightarrow C) = P(A \rightarrow C/B)$. This looks like a straightforward application of the definition, but it assumes that the conditional, (A→ C) expresses the same proposition in the prior context as it expresses in the posterior context. If this assumption is not satisfied, then the argument will suffer from a fallacy of equivocation. And it is clear, given the pragmatic constraint that our truth-conditional analysis imposed on the interpretation of indicative conditionals, that the invariance assumption will not hold. If what is accepted in a context changes, then different constraints are imposed on the interpretation of the indicative conditional, and this will require that in some cases, a different proposition is expressed in the posterior context.

Van Fraassen not only identified a questionable implicit premise of Lewis's proof of his triviality thesis, but also proved a counter-thesis—a possibility result. What Van Fraassen proved was this: for any probability model defined on a set of sentences of an extensional propositional language, one can extend it to a probability model for the richer language that adds a conditional connective, and in which Adams's thesis is satisfied. That is, for any two sentences A and B of the original language for which $P(A) \neq 0$, $P(A \rightarrow B) = P(B/A)$.

[11] Lewis 1976. See Hàjek and Hall 1994 for an excellent discussion of Lewis's result, and related issues.

[12] van Fraassen 1976. See also Stalnaker and Jeffrey 1994.

3. From More Fragile to More Robust Conditionals

So, one *can* give a truth-conditional form to an expressivist account of indicative conditionals—one that takes conditional speech acts as expressions of conditional belief, as represented by conditional credence. More generally, one can reconcile truth-conditional analyses of deontic and epistemic modal expressions with an account of their role in discourse that interprets them as doing something other than making factual claims. But one may ask—now focusing just on the case of conditionals—what advantage does this kind of formulation have? The response to the triviality arguments brings out the fragility of the propositions that are expressed, according to this formulation, by conditionals that conform to Adams's thesis. The model that van Fraassen constructed for his possibility proof is just one way of reconciling Adams's thesis with a truth-conditional analysis, but any way of constructing a model that does this job will have to define the model in terms of a given credence function, and when the credence function is changed, one will have to redefine the model if Adams's thesis is to hold also relative to the new credence function. And it is not just that the interpretation of the conditional must be highly context-dependent (that what proposition is expressed depends on context), but that the propositions themselves seem not to be detachable from the epistemic context in which they are expressed. That is, it is not clear how we can express, in a new context, after one's epistemic situation changes, what was said before with a conditional assertion.

The upshot is that the triviality results, and the response to them, provide a theoretical argument for the conclusion that any propositional analysis of conditionals that is compatible with Adams's thesis will necessarily be highly fragile: the proposition expressed by a conditional will shift with any shift in the epistemic state of the parties to the context. There are also intuitive examples that help to make this point, the most famous of which is Allan Gibbard's story about a poker game on a Mississippi riverboat:[13] Mr. Stone has raised, and it is up to Sly Pete to call or fold. Jack has seen both hands and knows that Pete has a losing hand. Zack has seen only Mr. Stone's hand, which is quite good, but he has signaled its content to Pete, so he knows that Pete will call only if he has a hand that can beat Mr. Stone's. At this point the room is cleared, and Jack and Zack convey their information to a third person, Allan. Jack says, in a note to Allan, "If Pete called, he lost," and Jack says, in a separate note, sent independently, "If Pete called, he won." Allan accepts both statements, and concludes that Peter folded.

The example has the following structure: There are three relevant alternative possibilities, X, Y, and Z. One person has information that excludes possibility Z, and a second person has information that excludes possibility Y. The two informants convey their information to a third person independently by making conditional assertions. What our truth-conditional formulation of the conditional assertion account says about the example is that the conditional statements made by the two informants must be interpreted relative to different contexts. Gibbard carefully sets

[13] Gibbard 1980.

up the example so that the *prior* context of each conditional assertion is the same: it is a context in which all three possibilities are compatible with what is presupposed. But our truth-conditional account treats the conditional assertion as a proposal to change the context to ensure that the sentence, as interpreted in the *posterior* context, is true. The context, as changed by Jack's statement, is different from the context, as changed by Zack's statement. Since the actual situation—the situation in which Pete knew he had a losing hand, and so folded—is compatible with what both Jack and Zack said, our truth-conditional account will say that both of their statements were true, which implied that the proposition Jack affirmed must be different from the one that Zack denied.

One might think that the point of giving a propositional analysis is to identify an item of information which can be detached from the context in which it is expressed, and from the epistemic situation of a person who believes it, an item that might be assessed, in retrospect, in a later context. But this does not seem possible with examples that fit the pattern of the Sly Pete story. Imagine Allan thinking to himself, after getting the messages from his informants (but still not having gotten definitive information about how the game ended), "Suppose Pete did call after all. Did he then win or lose? If he did, call then one of my informants was mistaken. Perhaps *Jack* was right, but Pete missed Zack's signal, and so mistakenly thought he had a winning hand. Or alternatively, perhaps Zack was right, but Jack, in observing the two hands, mistook a jack for a king, or a spade for a club, and so was wrong that Pete had a losing hand." It seems intuitively clear that Allan might think this way, even if he had complete confidence that both of his informants were correct in what they said. Suppose he takes himself to know, based on what his informants told him, that Peter folded. He can still consider what he would think if something he took himself to know turned out to be false. If this is the situation, then the question he is asking himself cannot be a question about the truth or falsity of propositions expressed by Jack and Zack. He is asking not "Which of my informants was wrong?" but rather, "*If* one of them is wrong (because, contrary to what I take myself to know, Pete did indeed call), which one of them is it?" So, Allan's conditional question is not a retrospective assessment of Jack's and Zack's statements. It is also clear that Allan's question is not the counterfactual question, "If Pete had called, would he have won or lost?" Assuming as we do that Allan trusts his informants, the answer to this question is clear: he would have lost. This does not show that Zack was wrong when he said, "If Pete called, he won" for the following reason: Zack was well aware that he was not in a position to rule out the possibility that Pete had a losing hand (in which case he would have lost if he had called), and yet he was in a position to say that if Pete called, he won. So, it can't be that his statement ruled out this possibility.

If conditionals of the kind that are involved in the Sly Pete story were the only ones, there would be little point in giving a truth-conditional formulation to the conditional assertion account, but with some indicative conditionals, and certainly with counterfactual conditionals, it does seem that we can identify something like an item of information that is not tied so tightly to the context in which it is expressed. In the Sly Pete story, there is no disagreement between Jack and Zack, even though their statements seem, on the surface, to make incompatible claims. It is just that each has only partial information, which when they share it with each other will change their epistemic situations so that the question of retrospective assessment of their apparently contrary claims does not arise. But the situation with other examples, including some that share the abstract structure of the Sly Pete story, will be different. Here is an old example I have discussed before, based on a case that Paul Grice described in his William James lectures in 1967:[14] There are three candidates: Wilson, Heath, and Thorpe. Alice is convinced that if the victor is not Wilson, it will be Heath, while Bert thinks she is wrong about this—if it is not Wilson who wins it will be Thorpe. The case is like the Sly Pete case in that there are three possibilities and two agents, one who takes herself to be in a position to rule out one of the three possibilities, while the other takes himself to be in a position to rule out a different possibility. But in contrast with the Sly Pete story, in this case the two hold their ground, each rejecting the other's claim. They take themselves to be disagreeing, and each may defend his or her claim against the other, presenting evidence that he or she is right. As they debate the issue, the context keeps shifting, both introducing new information that they take to support their case. The disagreement persists, and even though the context keeps changing as new information about polls, precedents, or policy positions is introduced, Alice and Bert take themselves to be addressing the same question throughout, and to be defending contrary answers to it. Even after the election, which Wilson wins, the disagreement may persist, with the conditional now taking a counterfactual form. Alice insists she was right—the election might have gone the other way, and if Wilson hadn't won, it would have been Heath. Bert agrees that the result might have been different, but remains convinced that if the winner hadn't been Wilson, it would have been Thorpe.

The fact of disagreement does not by itself require the conclusion that the conditional aims to make a factual claim that is true or false. Two people may disagree about what they are in a position to accept, or what beliefs it is reasonable for them to have, given their shared evidence, even if there is no factual claim that one regards as true and the other regards as false. But the dialectic of disagreement, and the retrospective assessment, in a later context (after more information has come) of what was said before, suggest that there is something like an item of information in dispute that persists through a shifting context, an item that a truth-condition would help to identify. Conditionals may begin simply as expressions of hedged commitments, reflecting holistic properties of one's current epistemic situation. Some conditional beliefs are ephemeral, reflecting parochial features of one's particular epistemic situation—what one happens to know and be ignorant of. The Sly Pete story was designed to be a case of this kind, a case where there is no retrospective question: which of the two apparently conflicting conditional claims—Jack's or Zack's, was correct. But in general, we have reason to try to find more stable or robust conditional beliefs that can be generalized and applied in a range of different particular epistemic situations. Sometimes it is appropriate to project one's epistemic priorities onto the world, to hypothesize that the world is such that one would be (under normal conditions) in a position to infer ψ upon learning ϕ. But in a particular case, it may be uncertain or controversial whether a dispute about a

[14] See Grice 1989, 64, and Stalnaker 1984, 113.

conditional is one that would be settled by a fact that is independent of the truth or falsity of the supposition or whether the disagreement just reflects a difference in the epistemic perspectives of the parties to the dispute. In the British election example, retrospective assessment after more of the relevant facts are in might reveal that the dispute about Wilson, Thorpe, and Heath was really more like the Sly Pete case, in that the right explanation for the disagreement was that Alice had information that, properly understood, was sufficient to rule out Thorpe, while Bert had information that sufficed to rule out Heath. But on the other hand, a retrospective assessment might reveal that there are factual questions on which their dispute turned that settle the question, if Wilson hadn't won, which of the other two would it have been?[15] Given that there may be, at the time of a dispute, uncertainty about whether there is a fact of the matter, and given that there is continuity between the more ephemeral conditionals that are essentially tied to a specific epistemic situation, and more robust conditionals that can be detached from their epistemic contexts and assessed against the facts, it is useful to have a semantic account of conditionals that covers the whole range of cases. A truth-conditional semantics that allows for both context-dependence and truth-value gaps is one that meets this condition.

4. Retrospective Assessment

The points made in the last section about the continuity between indicative and counterfactual conditionals, and about retrospective assessment of indicative conditional claims, are not matters of dispute between my account and Edgington's. In fact, she argues forcefully and persuasively for exactly this kind of continuity, using it as part of her argument for extending the non-truth-conditional analysis of indicative conditionals to the case of counterfactuals. She makes a convincing case that a counterfactual conditional may, in some cases, be understood as a restatement of what was said earlier, in a different context, with an indicative conditional. But there is a tension between this observation and Adams's thesis that conditional probabilities provide the assertability conditions for indicative conditionals. Let me first review one of her striking examples, and then say what I think the tension is.

According to Edgington's story,[16] our protagonist (call her Dorothy) is driving to the airport when her car breaks down, and as a result she misses her plane to Paris. This was fortunate, as it turned out, since the plane crashed and everyone on board was killed. Coincidentally, a fortuneteller had predicted this disaster, saying to Dorothy, "If you take that plane, you will be killed." Dorothy put no stock in the fortuneteller's prediction. She ignored it, and when she learned of the crash she rightly took it to be a ghastly coincidence. But still, when she learns of the crash,

[15] The question is not just whether there is a counterfactual question, but whether the answer to the counterfactual question counts as a retrospective assessment of the original epistemic dispute. As I argued above, in the Sly Pete story there is a clear correct answer to the counterfactual question, "If Pete had called, would he have won?" but it is not relevant to judging whether Jack or Zack were correct in their epistemic conditional judgments.

[16] Edgington 2004, 12.

she says of the fortuneteller, "My god, she was right! If I had taken that plane I would have been killed."

Edgington uses this story to make a number of points, but the one I want to focus on here is that what is said with the counterfactual is taken to be the same thing as what was said earlier by the fortuneteller with an indicative conditional. Edgington does not think that the claim that the fortuneteller was right is the claim that what she said was true, since on her account conditionals don't have truth-values, but she is observing that there is some kind of item that was expressed in the earlier context, and that can be expressed again, and assessed later, in a different context. This seems right to me, but it raises a problem for the hypothesis that the indicative conditional expresses a conditional belief, as represented by a conditional probability. The problem is that the appropriateness of the retrospective assessment of the indicative conditional suggests the following constraint on indicative conditionals—a kind of reflection principle: one should not make a conditional assertion if one has good reason to believe that a retrospective assessment would judge the assertion to have been mistaken. This constraint is not a problem for the fortuneteller example, where even though Dorothy later judges correctly that the indicative statement was right, neither she nor the fortuneteller had reason to think this in advance. Since there was at the time reason to think that the conditional claim would be seen in hindsight to be mistaken, the reflection principle implies, as one should expect, that the statement was unjustified. But there are other examples that show a conflict between what the conditional credence prescribes and what is required by the anticipation of a retrospective judgment. These include examples that have been posed in the literature as counterexamples to Adams's thesis. The counterexamples are controversial, but I want to suggest that Edgington's point about retrospective assessment provides some theoretical rationale for the judgment that the examples are indeed counterexamples.

Start with a story told by Vann McGee:[17] a contestant on a television game show ("To Tell the Truth") is believed to be Sherlock Holmes. It is not certain that he is Holmes—say that our observer is about 90 percent sure that he is. A man, Murdock, has died under suspicious circumstances, and before hearing what the game show contestant says, the observer was inclined to believe that the death was probably an accident. But the putative Holmes asserts that it was murder, and that he is almost certain that it was Brown who did it. He further asserts that it was definitely murder, so if it wasn't Brown who did it, it was someone else.

Since our subject is 90 percent sure that the contestant is indeed Holmes, and since he takes the perceptive Holmes to be authoritative on questions about the causes of suspicious deaths, he is highly confident that what the contestant said is true: it was murder, Brown did it, but if he didn't someone else did. Should he learn, however, that Brown did not do it, this would cast doubt on the judgment that the contestant is really Holmes, since it is unlikely that Holmes would be both confident and mistaken about who did it. But if the contestant is not the real Holmes, then there is no reason to believe that this is a case of murder at all.

[17] McGee 2000.

So, our subject's conditional credence—that someone else did it, on the condition that Brown did not—is much lower than his belief in the conditional asserted by the contestant—that if Brown didn't do it, someone else did.

If we apply the retrospective reflection principle to Holmes's conditional claim, we get some support for McGee's intuitive judgments about this example. Our subject is 90 percent certain that the contestant is Holmes. What he therefore expects (with 90 percent certainty) is that when the truth about the contestants is revealed at the end of the show, we will learn for sure that the contestant is indeed Holmes. If that happens, he will then have strong reason to believe that Murdoch's suspicious death was murder, that Brown did it, and that if he didn't, someone else did. At that point (if what our subject expects with 90 percent certainty does transpire), credence in the indicative conditional will coincide with the *revised* conditional credence, but at the earlier point, it is the *anticipated* conditional credence (what it is expected to be when the facts about the contestants comes in) rather than the conditional credence at that time that determines the credence in the conditional. Just to spell it out: let H be the proposition that the contestant is Holmes, B the proposition that Brown was the murderer, and E the proposition that someone else was the murderer. Let P be the subject's initial credence function and let P_H be the anticipated credence function—the one that the subject expects to have after learning H. Then while $P(E|{\sim}B)$ is not high, $P_H(E|{\sim}B)$ is high. Since the subject expects that he will have reason, in retrospect, to accept the conditional, he has reason to accept it in the initial state.

It should be noted that this example is not a case, like Edgington's plane crash case, where the retrospective judgment about the conditional is made with a counterfactual conditional. As McGee pointed out, in this case, there is a clear contrast between the indicative conditional statement (both at the earlier point, and at the point of retrospective assessment, after the identity of the contestant is revealed) and the corresponding counterfactual. Neither Holmes nor our subject is inclined to give any credence at all to the claim that if Brown hadn't done it, someone else would have. It might be a robust part of the theory about the crime that whoever did it acted alone, so that if the actual murderer (whoever it turns out to be) had not done it, the deed would not have been done. But there are other putative counterexamples to Adams's thesis where the anticipated retrospective judgment is made with a counterfactual conditional. The examples posed by Stefan Kaufmann[18] fit this pattern. Since these examples are more abstract than McGee's, involving balls drawn from urns, they also reveal more clearly the structure of the cases, and provide hard numbers for the probability judgments.

Suppose a ball is to be drawn at random from one of two urns, A or B. We are 90 percent certain that it will be from urn A, whose composition is as follows: forty-nine white balls, and one red ball with a black spot. Urn B also contains fifty balls, one white, and forty-nine red, all without a black spot. What is the right credence, given these facts about the case, for the conditional, if a red ball is drawn, it will have a black spot? There is some inclination to say that it is .9, since we are 90 percent sure that the urn is A, and a red ball will have a black spot if and only if it is drawn from urn A.

[18] Kaufmann 2004.

But the conditional probability of a black spot, given a red ball, is only about .155. For if a red ball is drawn, it is very probably from urn B, in which case it will lack the black spot.

What should one expect to happen in this case? It is almost 90 percent certain that a white ball will be drawn from urn A, in which case it will then be certain that if a red ball had been drawn (unlikely as that would have been), it would have had a black spot. If the counterfactual judgment is reasonably taken to be a retrospective judgment of the indicative conditional, then the person who claimed earlier that probably, if a red ball is drawn it will have a black spot will be vindicated. So since in the initial context, one gives high credence to this outcome, the reflection principle implies that one should at that point affirm this conditional (or judge it to be highly probable), and not the contrary one that goes by the conditional probability.

One's intuitions about this kind of case may be pulled in two directions. Some are inclined to think that one should still assess the indicative conditional by the conditional probability, and that intuitions to the contrary are based on a cognitive illusion. But there is more at stake here than just intuitions about examples. The putative counterexamples are all cases where the conditional credence is unstable in the sense that it would be easily changed, in one direction or the other, by salient new information, information that the agent knows would result in a more stable conditional probability judgment. The probability space is naturally partitioned into alternative hypotheses (the contestant is Holmes, or not; the urn is A, or B) that meet the following condition: Learning which of these alternative hypotheses is true would stabilize the conditional credence, and give it an objective rationale. One thus has reason to be interested in conditionals that express, not just the current disposition to change one's credences, but a more stable disposition to change one's credences that one *would* have if one learned certain salient facts. Such conditionals may help to give more structure to one's epistemic situation, and to identify epistemically valuable questions to ask in one's inquiries. The moves from indicative to subjunctive conditionals, and from subjective credence to objective chance are moves that aim to find robust reflections of epistemic priorities in the judgments of fact about the world.

The truth-conditional semantics provides a formal structure that helps to model this kind of move from conditionals that express local and ephemeral features of an epistemic situation to conditionals that aim to reflect robust features of the world. The pragmatic constraint on indicative conditionals gives a class of admissible selection functions, and in the limiting case—the most cautious and noncommittal kind of model—all admissible selection functions are on a par. In this kind of model, conditionals are true only if true relative to all admissible selection functions, and false only if false for all. Indicative conditionals, on this limiting interpretation, do no more than express the epistemic disposition to accept the consequent upon coming to accept the antecedent. The speech act is a proposal that this epistemic policy be adopted for the moment. But the admissibility constraint is also compatible with structure that further constrains the selection functions. The theory allows for a partition of the space of possibilities, with the selection function constrained to select a possible situation from within its partition cell. That is, if the partition is $X_1, \ldots X_n$, then it might be required that for any proposition ϕ that overlaps X_i, and for any

$w \in X_i$, $f(w,\phi) \in X_i$. This gives plausible models for McGee's "To tell the truth" example, and for the urn examples discussed by Kaufmann. The more fine-grained the partition, the more the conditional is making a factual claim on the world.[19]

The indicative conditional, because of the pragmatic admissibility constraint, remains epistemic, but where we find a more stable partition, it sometimes allows us to detach the factual claim implicit in the conditional judgment from uncertainty about the truth of the antecedent. The tense/aspect/mood morphology that we inaccurately label "subjunctive" signals that some of the presuppositions that define the speech context are being temporarily suspended, allowing us to say something like this: Independently of our actual knowledge or ignorance of the truth of the antecedent, the facts are such that (under normal conditions) if we didn't know whether the antecedent was true or not, we would be in a position to accept the consequent upon learning that the antecedent was true.

5. Chance and Counterfactuals

There seems to be no disagreement between Edgington's account of conditionals and the one I am promoting on the following two points: First, there is an item of some kind—something like a piece of information—that is expressed by a conditional, and that in some cases can be expressed and assessed in different contexts, including retrospective contexts in which what was earlier said with an indicative conditional is said with one that is counterfactual. Second, there is a distinction between judging such an item to be reasonable or justified and judging it to be *correct*. Edgington's plane crash example, with its prediction by the fortuneteller, was designed to provide a dramatic illustration of this distinction—a case where a speaker was completely unreasonable and unjustified in making a conditional assertion which nevertheless turned out to be correct. One might think that the essential point of the thesis that conditionals express propositions is the claim that they express items of information that can be detached from the context in which they are expressed, and assessed as correct or incorrect independently of whether believing them was reasonable or unreasonable. But Edgington aims to reconcile these points about conditional statements and beliefs with a non-truth-conditional account. She writes, "We are all familiar with the thought that rationally held beliefs may turn out false, and, conversely, something which there is no reason to believe may turn out true.... If that were my story [about the 'correct'/reasonable distinction in the case of the conditional], there would be no novelty or mystery. But that is not my story. Counterfactuals, like other conditionals, are believed to the extent that a certain conditional probability is judged to be high, and that is not the probability of the truth of a proposition."[20] In supporting her resistance to a truth-conditional analysis, she gives a variation of the plane crash example in which the fortuneteller's prediction was hedged: she said, "I'm pretty sure that if you fly this week you will be killed," and then it turned out to be 90 percent of the passengers, rather than all, who are

[19] Brian Skyrms developed, some time ago, formal machinery for modeling this kind of process. See Skyrms 1980 and 1984.

[20] Edgington 2004, 22.

killed in the crash of the plane. Dorothy's counterfactual judgment, made in retrospect, was that "it was very likely that I would have been killed, had I caught that plane." This retrospective probabilistic judgment is "right" even if there is no fact of the matter whether she would have been one of the 90 percent who were killed, or instead one of the 10 percent who survived. The conditional probability, she maintains, is not a probability of truth.

David Lewis's truth-conditional semantics for counterfactuals would judge that the counterfactual, "If I had caught the plane, I would have been killed," is determinately false in the case where there is only a 90 percent (or 99.9 percent) chance that the speaker would not have been among the few survivors. Edgington resists this conclusion; as she says, Lewis's truth conditions "make it too easy for a counterfactual to be plain false."[21] I agree with her about this, and I agree with her conclusion that we should judge the counterfactual to be highly probable, even though we may also say that there is no fact of the matter about whether it is true. As discussed above, the truth-conditional semantics I have defended allows for indeterminacy in application, with both indicative and subjunctive/counterfactual conditionals. So, our truth-conditional approach can allow that there is no fact of the matter about whether a counterfactual such as the one in the modified plane crash case is true or false. But we will have a satisfactory account of this kind of case, and a plausible reconciliation of a truth-conditional theory with Edgington's approach, only if we can agree that counterfactuals without truth-values can have well-defined probability values. In most familiar cases of semantic indeterminacy, and of a supervaluation strategy for modeling them, we don't get probability values for the indeterminate cases. One does not normally judge, of a borderline case of baldness, that the person is 70 percent likely to be bald. But there is a different kind of indeterminacy, at least according to some metaphysical views, where probability judgments of propositions that are thought to lack truth-value seem to be natural. Consider the metaphysical thesis that there is no fact of the matter about future contingents. On this picture, the past and present are settled, but the future is open, and all possible futures are on a par in the sense that there is (now) no fact of the matter about which of the alternative futures will be realized. Still, this metaphysical view allows one to make predictions, and to speculate about how things will turn out. A semantics for a language appropriate to this metaphysical theory will be based on a branching tree structure where the nodes of the tree represent possible present moments, and total paths through the tree represent possible histories. Though there is, at any time, no fact of the matter about which total path through the tree is the actual one, the semantics can assign truth-values to the sentences of the language relative to the total paths (possible histories), and one can do the compositional semantics on propositions that are defined as functions from histories to truth-values. So, the statement, "There will be a sea battle tomorrow" might be true relative to some total histories that share our present moment and false relative to others, and therefore (applying the supervaluation method) neither true nor false *simpliciter*. The temporalist metaphysics allows that one might make a speculative prediction—"There

[21] Ibid., 14.

will (I conjecture) be a sea battle tomorrow"—even when one knows that there is now no fact of the matter about whether one's prediction is true. Predictions are made to be assessed when the time comes.[22]

Now the key point for our purposes is that even if the proposition that there will be a sea battle tomorrow is now neither true nor false, that proposition might now have a probability value. Suppose, for example, that it is a fact about the present that there is a 75 percent chance of a sea battle. For this kind of more metaphysical indeterminacy, it is natural to take propositions that lack truth-values as having probability values. Is this probability of truth? It is probability of truth relative to a history.

You don't have to sign on to this metaphysical theory (as I do not) in order to find it intelligible (as I do) and to use it as a kind of precedent for a case where the thesis of metaphysical indeterminacy may be less controversial. Suppose we have a chance model, based on exactly the kind of tree structure used to model the story told just above. Attached to each branch from each choice point in the tree is a probability value—the chance, at that point, that that branch will be taken. The model is neutral about the metaphysical question just considered. One might be a metaphysical realist about the future, holding that there is a fact of the matter about which total history is actual, even though the fact is not determined by the present state of the world: on this realist account, future contingent propositions are true (now) in virtue of the way things will in the end turn out. Alternatively, one might sign on to the above temporalist metaphysical picture. Either way, each choice point in the model determines a class of histories, and one can define propositions as functions from histories to truth-values. On either metaphysical view, there will be a fact of the matter about what the objective probability of a proposition is at any given point. On the realist interpretation, a proposition about a future contingent might be both true, at a certain moment (because it is true relative to the history that will turn out to be actual), and also have (at that point) only a 10 percent chance of being true.

So, the semantics is metaphysically neutral. The temporalist and the realist both have reason to define propositions in terms of total histories, even though the temporalist thinks that there is no fact of the matter about which history is actual. In order to get the compositional semantics to work smoothly the temporalist takes the points of evaluation as something more fine-grained than what her metaphysics claims is determined by the facts, but making this move is metaphysically innocent, since she need not use any resources in her semantics that her metaphysics does not provide.

What I want to suggest is that we can make a similar move, this time to points of evaluation that are even more fine grained than total histories, in order to get a smooth compositional semantics for counterfactuals. Define a *choice function* as a function that determines not only a unique path through the tree, but also a choice at each choice point in the tree, including those on paths not taken.[23] Suppose we take our "propositions" to be functions, from total choice functions, rather than just total

[22] See Thomason 1970 for a semantics of this kind, and MacFarlane 2003 for a discussion of this kind of temporalist metaphysics.

[23] In addition, a choice function may order the choices at choice points with more than two alternatives.

histories, to truth-values.[24] This move, like the temporalist's, is metaphysically innocent, since all the material needed to define choice functions is available even if we deny, as seems plausible, that there is a fact of the matter about which choice function is the actual one. What this move buys us is this: first, there is a natural comparative similarity relation that orders the choice functions, relative to each choice function that can be used to define a selection function that will yield a unique nearest "possible world" (choice function) relative to propositions that are plausible antecedents for counterfactuals. Second, there is a natural probability distribution on the space of choice functions that will determine probability values for the counterfactuals. Counterfactuals that are true relative to some choice functions determined by a given history and false for others will have no truth-values, but well-defined probability values.

There is one further twist to be added to the story before we have completed what I hope is a reconciliation of Edgington's account with mine. We need to ensure that our chance structures can model a crucial distinction that Edgington emphasizes in her discussion of conditionals in hindsight—a distinction that involves the notion of causal independence. The temporal order implicit in the chance model already captures one constraint on the notion of causal or counterfactual independence: it is presupposed that earlier events are independent of later ones in the sense that if a later chance event had been different than it in fact was, the earlier one would still have been the same as it was. But this is a sufficient and non-necessary condition for causal or counterfactual independence. Edgington contrasts these two kinds of cases: (1) I pick a coin from a bowl of fair coins and flip it: it lands heads. If I had chosen a different coin instead, there is no reason to think it would have landed heads: the chance that that would have happened is 50/50. (2) Two coins are selected from the bowl and separately flipped, perhaps in different rooms, one slightly later than the other. Both land heads. This time it seems reasonable to say that if the coin that was flipped first had landed tails instead, the other coin would still have landed heads, as it actually did. The second kind of case has been labeled a Morgenbesser case, after an example attributed to Sydney Morgenbesser: I decline to bet on the flip of a coin, which is then flipped, landing heads. I say, "If I had bet on heads, I would have won," and this seems obviously right. It is also true that if I had bet on heads, at that point in time, there would have been only a 50/50 chance that the coin would land heads. But if, as is plausible, the flipping event was causally independent of the betting decision, the counterfactual statement will be correct. Morgenbesser cases, where later events are causally independent of some earlier ones, play a crucial role in Edgington's account of the hindsight judgments, where earlier indicative conditional claims are vindicated by later counterfactual judgments.

We can account for this distinction in the context of our tree-structure models by recognizing that distinct nodes of the tree may be linked, and that the only admissible choice functions are those that make corresponding choices at linked nodes. (I sketch

[24] See Thomason and Gupta 1980 for a semantic theory for conditionals and branching time that develops this strategy.

the details in an appendix.)[25] With this addition to the model, the natural comparative similarity relation specified, and the selection functions defined in terms of it, will get the right result, both about Morgenbesser cases, and about those which contrast with the Morgenbesser cases. And in a plausible model of this kind for Edgington's plane crash case, we will get the result that, even if there is no fact of the matter about whether Dorothy would have survived if she had made the plane, it is nevertheless true that she probably would not have survived (and that, in a sense, it is probably *true* that if she had made the plane, she would not have survived.)

We asked, at the start of section 2, what is the point of giving a truth-conditional form to the conditional assertion account? The answer was that there would perhaps be little point if all conditional assertions were the fragile kind that conform to Adams's thesis, and that reflect only ephemeral features of a local epistemic situation, but that the truth-conditional form facilitated an account of the continuity between these conditionals and more robust ones, including counterfactuals—conditionals that seem to reflect factual judgments about the objective world. This answer works only if we have reason to give a truth-conditional account of the more robust conditionals that seem to express some kind of items of information that can be detached from the contexts in which they are expressed, and assessed as correct or not, right or wrong, but it is easier to justify a truth-conditional account of the more robust conditionals. It is not only that it seems an intuitive strain to withhold the words "true" and "false" in the cases where we judge the counterfactual to be "right" or "wrong," "correct" or "incorrect." It is that these items of information (unlike the more ephemeral conditional assertions) combine naturally with each other, and with statements and clauses that uncontroversially have truth conditions. The traditional defense of a truth-conditional approach—that it gets the compositional semantics to work—is an important part of the rationale.

Edgington has used the slogan "objectivity without truth" to sum up her approach to counterfactuals. I was struck by the juxtaposition of this slogan with the title of a book by Max Kölbel defending relativistic semantics, *Truth without Objectivity*.[26] The first slogan is represented by statements like "She would have been killed if she had made the plane" in the revised version of Edgington's example: the statement was judged to be objectively probable, but neither true nor false. The second slogan fits with a semantics that assigns truth-values to points of evaluation (such as choice functions) that do not represent something there is an objective fact of the matter about. I think we need both objectivity without truth, and truth without objectivity for a satisfactory account of conditionals.

Appendix

I will give a very rough sketch of a simple theory for modeling chance processes. I borrow some resources from game theory to define the models.

[25] Thomason and Gupta considered a Morgenbesser case (though not under this name), and the way the choice functions need to be restricted to account for such cases. See Thomason and Gupta 1980, 82–4.

[26] Kölbel 2002.

A game-theoretic representation of an extensive form game is a tree structure with the branches of the tree representing the choices available to one of the players at the point in the game represented by the node of the tree from which the branches branch. The nodes are labeled by the player who controls the choice at that node. In games involving chance events, one of the "players" is named "chance" or "nature," and for this player's choice points, the branches are assigned probabilities.

In so-called "games of imperfect information," distinct nodes of the tree may correspond to a single choice point. Suppose, for example, player 2's choice of paper, scissors, or rock is temporally later than player 1's choice of one of these options, but the choices must be made independently—as if simultaneously. In this case, player 2's choice point corresponds to three different nodes of the tree, the ones that result from each of player 1's possible choices. In the game-theoretic representation, the three nodes that follow player 1's choice are said to be *informationally equivalent*, and the set of the three equivalent nodes is an *information set*. The notion of an information set is usually explained intuitively in epistemic terms: player 2 must choose in ignorance of player 1's choice. But the distinction concerns the causal structure of the game, and is not essentially epistemic.[27] Player 2's choice is informationally independent of player 1's choice if and only if the choices are causally independent, which implies that if player 1 had chosen differently, player 2's choice would still have been the same. So informational equivalence can apply to chance moves (where epistemic notions don't apply, since the "player" is not an agent), as well as to moves of rational agents. So, suppose we have a game that involves the kind of situation described in the Morgenbesser examples discussed in the text. At the root node of the tree a player chooses whether to bet heads on a coin flip or to decline the bet, leading to two nodes that result from the two possible choices. Then the coin is flipped, independently of whether the bet was accepted. If the player declined the bet, and the coin then lands heads, the player can truly say that if he had accepted the bet, he would have won. The difference between the case where the result of the flip is causally independent of the prior choice and the case where it is not is represented in the model by difference between a case where the two nodes that represent the alternative choices are informationally equivalent and a case where they are not.

A pure chance model is a special case of a game—one that really takes us out of the realm of game theory, since there are no real players, just chance. We have the same kind of tree structure, with information sets in cases of causally independent chance events. Since every branch starts from a node "controlled" by chance, every branch is assigned a probability value.

Here is a formal definition of this kind of structure.[28] (To avoid technical complications, I will restrict attention to finite models.)

[27] One problem with the epistemic explanation of informational equivalence is this: it is sometimes assumed that players know that all players act rationally, and also know what the payoffs for all the players are. In a game (such as a one-shot prisoners' dilemma) where players make their choices independently, but where each player has a dominating choice, these assumptions will imply that player 2 knows what player 1 did, even though player 2's choice is informationally independent of player 1's choice.

[28] See Osborne and Rubinstein 1994, 200, for a definition of an extensive form game with imperfect information that defines the tree structure in this way.

A "chance model" is a structure ⟨H, c⟩, where H is a finite set of finite sequences meeting these conditions:

(1) $0 \in H$
(2) If $(a1 \ldots am) \in H$, and $k < m$, then $(a1 \ldots ak) \in H$.
(3) If $(a1 \ldots am) \in H$ and $i \neq j$, then $ai \neq aj$

This definition determines a finite tree structure. The elements of H may be thought of either as the nodes of the tree (with 0 the root node), or as all the paths through the tree up to a certain point (since there is a one-one correspondence between the nodes of the tree and the paths from the root to that node). The elements of the sequences are the branches of the tree; they represent *outcomes* of the chance event that occurs, or that might occur, at that point in the process. We define Z—the set of *terminal nodes* or *complete histories*—as those $h \in H$ for which there is no a such that $(h,a) \in H$. For each non-terminal node, we can define the branches from that node as follows: $A(h) = \{a: (h,a) \in H\}$.

Condition (3) above prohibits an element from occurring more than once on any given path. An element may occur more than once on the tree, but we add the following condition that constrains how this may occur:

(4) Either $A(hi) = A(hj)$ or $A(hi) \cap A(hj)$ is empty.

Two nodes h_i and h_j are defined as *informational equivalent* if and only if $A(hi) = A(hj)$.

The second element of the structure that defines the model, c, is a function assigning probability values to each branch of the tree. So, for any node h, c(h) will be a real number in the interval (0,1). For any nonterminal node h, the values of c for the members of A(h) must sum to one.

In the general case of a model of an extensive form game, one can define *strategies* for each of the players, which are functions taking each of that player's choice points to a choice. A strategy profile is a sequence of strategies, one for each player, which determines not only a complete path through the tree, but also answers to questions about what players would have done if choice points that were not reached had, contrary to fact, been reached. We can define an analogue of a strategy profile for our pure chance model: A *choice function* is a function taking each choice point to one of the choices available at that point. We may also define an *extended choice function* as a function that in addition orders the alternative available at choice points where there are more than two available options.

Now we can take a possible world to be a complete path through the tree, but we might also use a more fine-grained abstract object to be the points of evaluation in a possible worlds model: complete choice functions, or extended choice functions. Our model determines a probability function for the space of possibilities, defined in this way. To determine the weight to be assigned, in the probability space, to a given choice function, just take the product of the probabilities of each choice at each choice point.

Suppose we take choice functions as the points of evaluation in our semantics—the "possible worlds." There is then a natural comparative similarity relation that can be used to define a selection function for interpreting conditionals. The idea is simple: first, maximize agreement on all *reached* choice points; second, among choice functions that are tied with respect to this condition, maximize agreement on all choice points. These conditions will not yield a total ordering of all choice functions, relative to each choice function, but it will suffice to determine a unique closest possible world for many potential suppositions, including all where the proposition supposed specifies a particular counterfactual event.

Here are two contrasting examples: (1) there are two urns, one with nine red balls and one black, the other with nine black and one red. First a fair coin is flipped, and if it is heads, a ball is selected at random from the first urn, and if is tails from the second urn. The coin landed heads, and a black ball was drawn. What would have happened if the coin had landed tails? Our "actual" choice function will determine a choice at the unreached choice point where a ball is drawn from the second urn, and this choice will be preserved in the "closest" choice function to the "actual" one. (2) Two coins are flipped independently, the first landing heads, the second tails. Since the flips were independent, the result of either would have been the same if the other had been different, and our model gives this result. But suppose our counterfactual supposition is that the two coins land the same way (either both tails or both heads). Which would it have been? Our account gives no answer to this question, even given a full choice function, which seems intuitively right. This is just like Quine's notorious question whether if Bizet and Verdi had been compatriots, they would have been French or Italian. In this kind of case, we not only should not expect a truth-value for the counterfactual, we also should not expect a probability value.

For those cases where the semantics yields a determinate truth-value for a conditional (relative to each choice function), it will also yield a determinate probability value, relative to each node of the model.

It is, of course, a fiction that the choice functions are the possible worlds if this requires that there be a fact of the matter about which choice function is the actual one. But we can say that conditionals (and propositions generally) are true relative to a given history (represented by a path through the tree) if and only if true for all choice functions that determine that path. Some counterfactuals will lack truth-values in the supervaluation, but they still will have well-defined probability values, determined by the chance model, and the path through its tree.[29]

[29] Thanks to Lee Walters for very helpful comments on a draft of this chapter.

12

Counterfactuals and Humean Reduction

1. Introduction

David Lewis started his career with two major constructive projects, each aiming to do something that his teachers had argued cannot be done. First, in his dissertation and the book *Convention* that grew out of it, he argued that, despite what W. V. Quine believed, one *can* make sense of truth by convention. Then, in his book *Counterfactuals* he argued that, despite what Nelson Goodman had reluctantly concluded, one *can* give a reductive analysis of counterfactual conditionals. My topic is this second project: I will try to spell out the ways in which Lewis's aim was a reductive analysis, and the ways that it compares and contrasts with Goodman's project. It's not that Lewis argued that one could do exactly what Goodman gave up on doing; the framework in which he set the problem up was strikingly different. But both Goodman and Lewis were animated by Humean skepticism about natural necessity. What needed to be analyzed away was a family of concepts that apparently described or implied real relations and connections between distinct events. For both Goodman and Lewis, causal dependence and independence, capacities, dispositions, potentialities, and propensities were problematic, and so in need of analysis. But pinning any project of analysis down requires getting clear about the linguistic and conceptual resources that are available—about the unproblematic base to which the family of problematic concepts was to be reduced. A reductive project may fail, not just because no satisfactory analysis can be found, but also because the distinction between the problematic concepts and the unproblematic base is not sufficiently clear or well motivated. I will argue that Lewis's reductive project does not succeed, partly for this kind of reason. If I am right, this raises the same question that the acknowledged failure of Goodman's project raised: what is the alternative to a reductive analysis? If we conclude that no reductive analysis is to be had, but still find a concept or family of concepts philosophically problematic—in need of some kind of explanation—what do we do?

Here is my plan: I will start, in section 2, with Goodman's project, saying what his aim was, spelling out the resources that he allowed himself to use in order to accomplish this aim, and explaining why the aim could not be accomplished with these resources. I will conclude this section by considering Goodman's response to the acknowledged failure of his initial attempt at analysis: how he proposed to redefine his Humean project.[1] In section 3, I will sketch Lewis's project, which has two parts: an abstract formal semantic analysis of conditionals, and an attempt to explain the primitive parameters of the semantic models in a way that is austere enough to satisfy the Humean strictures, but rich enough to provide an intuitively satisfactory account of counterfactual conditionals, and other concepts in the family that the Humean argues are in need of explanation. In the next two sections I will look more closely and critically at the conceptual resources that Lewis used to characterize the base to which counterfactuals, laws of nature, causation, dispositions, capacities, propensities, and potentialities were to be reduced: in section 4 at the thesis of Humean supervenience, and in section 5 at Lewis's notion of a natural property. I will argue that Lewis does not succeed in identifying a suitable base for a reduction of counterfactuals and other causal notions, and then conclude, in section 6, with a brief remark on the problems that remain if we reject the project of Humean reduction.

2. Goodman's Project

One might think that if certain concepts are problematic, then there is some problem about them. What are the problems about dispositions, counterfactuals, and possibilities that, according to Goodman, "are among the most urgent and most pervasive that confront us today in the theory of knowledge and the philosophy of science"? In a famous passage in *Fact, Fiction and Forecast*, Goodman gave the following non-answer to this question:

> A philosophical problem is a call to provide an adequate explanation in terms of an acceptable basis.... What intrigues us as a problem, and what will satisfy us as a solution, will depend upon the line we draw between what is already clear and what needs to be clarified.... In the absence of any convenient and reliable criterion of what is clear, the individual thinker can only search his philosophical conscience. As is the way with conscience, it is elusive, variable, and too easily silenced in the face of hardship or temptation. At best it yields only specific judgments.... Indeed this talk of conscience is simply a figurative way of disclaiming any idea of justifying these basic judgments. (31–2)

If the kind of explanation one gives is a reductive analysis, then one does not need to say what the problem is in order to solve it. Whatever is problematic about some concept, the problem will disappear if one gives an eliminative definition of the concept in terms of notions that do not have that problem, whatever it is. But if the project of reductive analysis fails (as it usually does) then one will need to identify the problem or problems in order to see what kind of explanation, short of reduction, might solve it or them. Identifying the problems might also help to make clear what will count as an acceptable basis for the philosophical explanation. The terms in which one explains some problematic concept must be clear with respect to the particular problems at issue, but an analysis or explanation might still be successful,

[1] All Goodman quotations in this section are from Goodman 1983, with page references in parentheses. The first edition was published in 1953, and chapter 1 is a paper discussing the problem of analyzing counterfactuals, first published in 1946.

even if the notions one starts with are themselves problematic, but face different problems. (For example, one might think that the concepts of truth, belief, and justification all face serious philosophical problems, but still think that if the traditional justified true belief analysis of knowledge had not fallen to counterexamples, it would have provided a solution to problems about skepticism.)

It is clear enough why *Hume* thought that notions involving natural necessity were problematic and in need of analysis. All legitimate ideas are copies of, or at least analyzable in terms of, sense impressions, and there is no sense impression corresponding to the idea of causation. Goodman's worries about counterfactuals were not tied to Hume's specific doctrine about the empirical basis for concepts, or to the verificationist doctrine that was its twentieth-century descendant, but they are worries that have their source in empiricism. The logical empiricist project of explaining theoretical scientific notions in terms of more directly observational notions was a dauntingly difficult one, but it was evident to those pursuing this project that it would be a lot easier if one were allowed to include a counterfactual conditional operator among the logical resources used for such analyses (one of the temptations that the puritanical Goodman quotation above warns us to resist). Strictly extensional logical resources were unproblematic. Intensional connectives and operators threatened to smuggle nonempirical content into the concepts analyzed in terms of them, but if one could give truth-conditions for sentences involving such a connective, using only extensional logical resources, that would justify using counterfactuals in one's explanations of the relations between theory and observation. The general form of analysis that Goodman started with was this: a counterfactual conditional, $(\phi \Box\!\!\rightarrow \psi)$, is true if and only if ψ can be deduced from ϕ, conjoined with a true set of laws of nature and with a suitable set of true sentences (the relevant conditions). Filling out this skeleton of an analysis required two things: first, an account of laws of nature, distinguishing them from mere accidental generalizations, and second, a distinction between suitable and unsuitable factual truths—between those truths permitted as premises in the derivation of the consequent and those that are not. Goodman considered the two problems separately, but he might have put them together into the single problem of identifying the contingent truths that are suitable as additional premises in a derivation of the consequent from the antecedent. It was obvious from the start that one couldn't allow all truths to be premises without trivializing the analysis, since in the case of counterfactuals, the negation of the antecedent is true, and so a set of all truths plus the antecedent will be inconsistent, making every counterfactual true. Goodman tried on various restrictions: the relevant conditions must be compatible with the antecedent, compatible with both the consequent and the negation of the consequent, etc. It was also proposed to add a negative condition—that there *not* be a suitable set of premises which, when conjoined with the antecedent was sufficient to derive the negation of the consequent. Each attempt that Goodman considered could be trivialized by showing that it either made all counterfactuals true, or none of them. Then, as a rhetorical move preliminary to giving up, Goodman proposed that the set of relevant conditions be required to be not only consistent with, but *cotenable* with the antecedent, where a sentence is cotenable with the antecedent if and only if it is not the case that it would be false if the antecedent were true. But as he then pointed out, with this analysis, "[We] find ourselves involved in an infinite regression or a circle, for cotenability is defined in terms of counterfactuals, yet the meaning of counterfactuals is defined in terms of cotenability." (16) At this point, he gave up. Six years after the publication of his initial paper, he was still smarting from "the unsullied record of frustration" and the "years of beating our heads against the same wall and of chasing eagerly up the same blind alleys," (38) and so was ready to try something different. Before looking at the new strategy that he proposed, let me make two general remarks about the failed project.

First, Goodman did not give an argument that his project *had* to fail. He just showed that none of the particular proposals that he tried were successful. But he could have given what is close to a proof that the resources he allowed himself were bound to be insufficient. In specifying relevant conditions (the class of sentences that were admissible as premises to be added to the antecedent in a derivation of the consequent), one is allowed to appeal only to (1) the truth-value of the sentences (only truths were allowed) and (2) the logical relations between the candidate sentence and the antecedent and consequent sentences. One could require that to be a relevant condition, the candidate sentence had to be compatible or incompatible with or, entail or be entailed by the antecedent, or the consequent, or their negations. But it was not permissible to appeal to the *syntactic* structure of the sentences (for example, whether they were atomic, disjunctions, or negations), since that is an arbitrary matter of formulation. Using a modeling tool that Goodman would not have found congenial, but which is still suitable for picturing the information he allowed himself for his reductive project, we can represent the situation with a Venn diagram with just two circles, one for the antecedent A and one for the consequent C, and a point α to represent the actual world (Figure 12.1).

All of the four regions distinguished in the diagram will be open, since we assume that the antecedent and consequent are logically independent. The point α is outside of both the A and C circles, since we assume that A and C are both in fact false. The diagram then contains all of the information that might be used to decide whether any arbitrary third circle is an admissible relevant condition, and so the diagram contains all of the information that might be used to determine whether the counterfactual $(A \Box\!\!\rightarrow C)$ is true or false in the actual situation, α. But obviously, the diagram is the same for any two sentences A and C that are both false, and logically independent. Without adding some additional structure, there is nothing that could distinguish true from false counterfactuals.

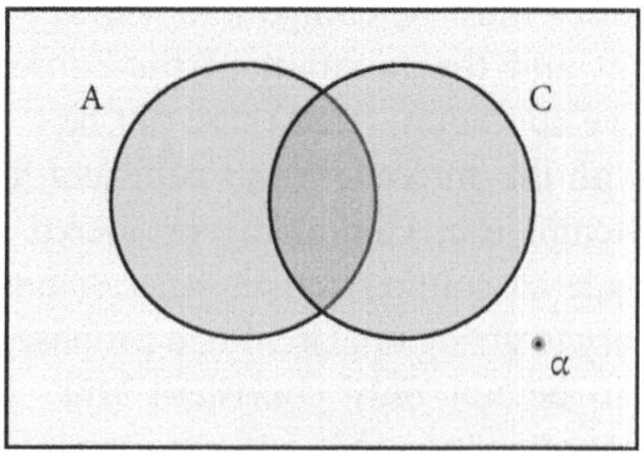

Figure 12.1

The second point to note is that Goodman might have paused after offering his circular cotenability analysis, but before turning away, in order to consider the consequences that this proposal has for the logical structure of counterfactuals. While failing to be a reductive analysis, the proposal is not thereby empty. Suppose one were to take the counterfactual to be primitive, and interpret the two parts of the circular analysis (cotenability defined in terms of the counterfactual, and the counterfactual defined in terms of cotenability) as postulates. There are a few details to pin down to make this suggestion precise, but on a straightforward way of spelling it out, the result would be a structure that is essentially the same as that of the kind of abstract semantics that was first proposed more than twenty years after Goodman floated the cotenability proposal. Fixation on the general philosophical strategy of reductive analysis led Goodman, and everyone else at the time, to ignore any consideration of a line of inquiry that later had interesting results, and helped to sharpen Goodman's problem, if not to solve it.

So, the exploration of the logical structure of the counterfactual was not part of Goodman's new direction. What was the strategy that he proposed to adopt after the failure of the initial project? His suggestion was that we give up on counterfactuals and start instead with dispositions.

> In dealing with counterfactuals . . . we are expressly concerning ourselves with a *form* of a statement, and the pattern of analysis we see is largely dictated by the structure of the conditional. This structure, although it promised at the outset to be a valuable aid, may actually have become a hindrance. The very disanalysis effected by returning to consider dispositional statements, which are indicative and simple in form, may free us to explore a better scheme of analysis. . . . I suspect that the problem of dispositions is really simpler than the problem of counterfactuals. (38–9)

Although it is clear that Goodman's new approach involves a switch in focus from counterfactuals to dispositions, it is less clear exactly what kind of analysis of them his new strategy requires. He suggests that we have already defused a part of the problem, simply by noting that "dispositional as well as manifest predicates are labels used in classifying *actual* things" (59). They require no departure from standard extensional logic: they are ordinary predicates with ordinary extensions. He also notes the pervasiveness of dispositional predicates: adjectives like "hard" and "red" are as dispositional as words with "a tell-tale suffix like 'ible' or 'able'. Indeed, almost every predicate commonly thought of as describing a lasting objective characteristic of a thing is as much a dispositional predicate as any other" (41–2) Dispositions remain problematic since "they seem to be applied to things in virtue of possible rather than actual occurrences" (42), but there are no merely possible occurrences. So, the problem is "to explain how dispositional predicates can be assigned to things solely on the basis of actual occurrences" (42). He goes on to suggest that the move from a manifest predicate such as "flexes" to a dispositional predicate such as "flexible" involves a kind of conceptual projection. Things not under suitable pressure neither flex nor fail to flex, but we can extend the flex/fail-to-flex distinction to a wider range of cases. That is, the extension of the new predicate "flexible" coincides with the extension of "flexes" within the domain of things under suitable pressure, but also may be applied to things outside of that domain. We extend the distinction, Goodman suggests, by using the same principles we use to make *inductive* projections: "The problem of dispositions looks suspiciously like one of the philosopher's oldest friends and enemies: the problem of induction. Indeed, the two are but different aspects of the general problem of proceeding from a given set of cases to a wider set. The critical questions throughout are the same: when, how, why is such a transition or expansion legitimate?" (58). Goodman's attention then shifts to the problem of induction, and the task of constructing a substantive theory of confirmation. Counterfactuals and dispositions fade into the background and are not mentioned again.

Goodman started with an uncompromising line: certain notions are clear, others in need of explanation. The family of concepts that Hume found suspect—potentialities, capacities, and dispositions, causal dependence and independence, objective chance—are central to our science and epistemology, but this is reason to seek clarification, not to succumb to the temptation of taking them to be clear. They (according to his philosophical conscience) fall on the wrong side of a bright line. Furthermore, Goodman suggested that philosophical explanation requires analysis, by which he meant reductive definition. He was scornful of the idea of a partial definition: concepts are either fully defined, or allowed as primitives (see pp. 46–7). But by the end of the discussion of dispositions, there is at least some suggestion that the problem has been defused, not by analysis, but by explaining why it is acceptable to take some dispositional predicates as primitives. After all, if the simple paradigm cases of observational terms like "red" (not of a phenomenalist language, but of a common-sense physical-thing language) are dispositional, then why should we think that dispositional predicates are necessarily problematic? We have to explain how we can have evidence that such predicates apply to things, even in circumstances where the disposition is not manifested, but that is a general epistemological problem. We have, Goodman argued, gotten rid of reference to merely possible things and occurrences, and perhaps that is enough to make at least some dispositions, with their limited counterfactual consequences, acceptable as they are. Compare Quine's discussion of counterfactuals and dispositions, which was heavily influenced by Goodman's ideas. Quine says there that "the subjunctive conditional has no place in an austere canonical notation for science," but that "we remain free to allow ourselves one by one any general terms we like, however subjunctive or dispositional their explanations."[2]

3. Lewis's Project

As I noted, Goodman's circular cotenability analysis contained the seeds of an account of the abstract logical structure of counterfactual conditionals, but he did not stop to develop that account. Instead, he concluded that the focus on logical structure was a hindrance rather than a help. Lewis's project begins by returning the focus to the compositional semantic structure of conditionals. He constructs a formal language with primitive conditional connectives, and a semantics for interpreting

[2] Quine 1960, 225.

them developed in the framework of possible worlds—a framework that did not exist when Goodman first attacked the problem of counterfactuals, and that he found extremely uncongenial after it was developed. Lewis's semantics gave truth-conditions for the interdefinable 'would' and 'might' conditionals in terms of a parameter of a model—a three-place relation of comparative similarity between possible worlds (y is more similar to x than z is to x). The rough idea is that a conditional, ($\phi \mathrel{\Box\!\!\rightarrow} \psi$) is true in a world x if and only if ψ is true in all of the ϕ-worlds that are most similar to x.[3] The formal properties of any comparative similarity relation will help to explain the logical properties of the conditionals whose truth conditions are given in terms of it. For example, since we can assume (no matter what the relevant respects of similarity are) that there is no y more similar to x than x is to itself, the analysis will validate modus ponens.

Although it imposes and explains a logical structure, Lewis's abstract semantics does not, by itself, provide a Humean reduction, since truth conditions are given relative to a primitive parameter of the interpretation—the comparative similarity relation. The formal properties of a relation of this kind are not enough to identify it. Unless and until this relation is explained in a way that shows it to be unproblematic in the relevant sense, one has not solved the problem, but only given a framework that sharpens it by separating the logical problem of compositional structure from the substantive problem of specifying the respects of similarity that are relevant to the interpretation of counterfactuals, and so to other concepts that might be defined in terms of counterfactuals. So, the substantive problem is where the action is, if a Humean reduction is to be successful, and what is required is not only a counterexample-free specification of the relevant respects of similarity between worlds, but also a clear and well-motivated identification of the resources that are admissible to the basis for the reduction. I think Lewis would agree with this assessment, and I will consider below his strategy for addressing the substantive problem.

In my own early paper on conditionals, which developed and defended a semantic analysis similar to Lewis's in the same possible-worlds framework, I distinguished the logical problem from the substantive problem (there, perhaps misleadingly, labeled "the pragmatic problem"[4]). While I thought there was more to be said about the substantive constraints on the selection function that encoded a comparative similarity relation, I did not think (then or later) that the problem was to find a reduction to concepts on some more basic level. And I argued that even without saying more about further constraints on comparative similarity, the abstract analysis does help to defuse one philosophical problem about counterfactuals—what I called "the

[3] This is only roughly right because it doesn't cover the case where there are no closest ϕ-worlds to x because there is an infinite sequence of ϕ-worlds that are closer and closer to x, with no last term. In this case, the conditional will be true in x if and only ψ is true at some ϕ-world, y, and also at all ϕ-worlds that are at least as similar to x as y is.

[4] The label, and my discussion in that paper, may be misleading because they blur the line between two different distinctions: (1) between abstract formal constraints on a relation of comparative similarity and substantive constraints on the relevant respects of similarity, and (2) between constraints that are part of the semantics of the conditional and constraints that are determined by the context of use of the conditional.

epistemological problem"—a problem that may be part of the reason that the empiricist finds counterfactuals and related notions problematic. As I put the problem then: "Counterfactuals seem to be . . . contingent, statements about unrealized possibilities. But contingent statements must be capable of confirmation by empirical evidence, and the investigator can gather evidence only in the actual world. How are conditionals that are both empirical and contrary-to-fact possible at all?" This problem is similar to a worry expressed by Goodman about dispositions: "The peculiarity of dispositional predicates is that they seem to be applied to things in virtue of possible rather than actual occurrences" (Goodman 1983, 42). This was my suggestion for defusing the problem: "There is no mystery to the fact that I can partially define a possible world in such a way that I am ignorant of some of the determinate truths in that world." This will happen if I define the world as a function of the actual world—in particular as similar to the actual world in some respect. For example, if I were to stipulate that the possible worlds I want to consider are those in which a certain yacht was two feet longer than it actually is, then it is easy to see how empirical evidence about the actual length of the yacht is relevant to what is true in this counterfactual world. If an abstract formal semantics of the kind that both I and Lewis proposed is right, then we can see that "conditionals do, implicitly and by convention, what is done explicitly and by stipulation" in an example such as the one about the yacht (Stalnaker 1968, 99.) But even if this succeeds in defusing one general epistemological worry about counterfactuals and dispositions, there remains the problem of saying more, and of saying what more needs to be said, about the particular ways in which the selected counterfactual possibilities are similar to the world at which the counterfactual conditional is being evaluated.

In introducing his semantic analysis of counterfactuals, Lewis noted that the truth conditions he was going to propose were stated in terms of a primitive parameter of a model for the language, to be fixed within rough limits by the context of use, but he claimed that this parameter is a familiar one that had application that was independent of its use in the interpretation of counterfactuals. The claim that this parameter—the relation of comparative similarity—was familiar and independent of its use in the analysis of counterfactuals was supported by an example of an intuitive judgment of comparative similarity between cities ("Seattle resembles San Francisco more closely than it resembles Los Angeles"). He granted that possible worlds are usually bigger than cities, and may differ in a wider variety of ways, but "still, any problems posed by my use of comparative similarity differ only in degree, not in kind, from problems about similarity that we would be stuck with no matter what we did about counterfactuals. Somehow we *do* have a familiar notion of comparative overall similarity, even of comparative similarity of big, complicated variegated things" (Lewis 1973, 92).

There are two problems with this reliance on the familiarity of the intuitive idea of overall similarity; one of them has been often noted, but the other is, I will suggest, the more serious problem, if the analysis is to be understood as part of a reductive account of notions that are problematic from a Humean point of view. The first problem is that if the notion of comparative similarity between worlds that is relevant to the interpretation of counterfactuals is the familiar one that we are presupposing when we judge that Lewis's claim about Seattle, San Francisco, and Los Angeles seems clearly true, then the analysis will face many counterexamples. Take any case where it seems intuitively that a certain choice not taken would have changed the course of history in dramatic ways if it had been taken. Possible worlds where the counterfactual choice is made, but the dramatic changes do not take place and things somehow manage to proceed in pretty much the way they actually did will seem, intuitively, to be more similar to the actual world. So, if the truth of the counterfactuals depends on what happens in the most similar worlds (in this intuitive sense), the intuitively true counterfactuals will come out false (see Fine 1975). Lewis responded to this problem by saying that he never meant to suggest that the particular respects of similarity that are in play in the example of the cities are the ones that are relevant to counterfactuals. The point of the example of the cities, he said, was just to respond to a general skepticism about the notion of similarity (see Lewis 1986a, 52–5). In later work, Lewis made a detailed proposal about the respects of similarity that are relevant to the interpretation of counterfactuals and he expressed surprise that he was interpreted as changing his account, rather than filling in more details that are relevant to one standard kind of context. But the discussion of intuitions about overall similarity in the book did seem to put more weight on that notion than it can bear.

The more serious problem with the reliance on some more or less intuitive notion of similarity is that it fails to reveal whether or not the respects of similarity that we are implicitly relying on include similarity with respect to features of the world that are suspect from a Humean point of view. One might think that an analysis in terms of similarity would be acceptable since the Humean is not skeptical about either purely spatial-temporal relations or relations of similarity and difference, which can be explained in terms of the sharing of properties. But the acceptability of a similarity relation depends on the acceptability of the properties that are the relevant respects of similarity. One would, of course, not have a Humean reduction if the properties shared by things that are similar in the relevant respect were powers, dispositions, capacities, and potentialities. Lewis's more detailed specification of the relevant respects of similarity aims to address this problem, as well as to avoid counterexamples, but doing so requires not only spelling out particular criteria for comparative similarity, but also showing how to describe the things to be compared—those "big, complicated, variegated things," possible worlds—in a way that is untainted by the problematic notions. That requires a general metaphysical theory, but of course Lewis has one.

4. Humean Supervenience

Lewis characterizes the general thesis that motivates his reductionist program this way:

> Humean supervenience is named in honor of the great denier of necessary connections. It is the doctrine that all there is to the world is a vast mosaic of local matters of particular fact, just one little thing after another.... We have geometry: a system of external relations of spatio-temporal distance between points, maybe points of space-time itself, maybe point-sized bits of matter or aether or fields, maybe both. And at those points we have local qualities.... And that

is all. There is no difference without a difference in the arrangement of qualities. All else supervenes on that.[5]

Lewis holds that Humean supervenience is a contingent thesis, and he takes his task not to be to establish its truth, but to rebut objections to it by philosophers who "claim that one or another commonplace feature of the world cannot supervene on the arrangement of qualities.... Being a commonsensical fellow (except where unactualized possible worlds are concerned) I will seldom deny that the features in question exist. I grant their existence and do my best to show how they can, after all, supervene on the arrangement of qualities."[6] The plan for doing this follows a careful order, to avoid circularity: first, Lewis offers an account of laws of nature, following an old suggestion of Frank Ramsey that the laws are those universal generalizations "that achieve an unexcelled combination of simplicity and strength." The "best system" analysis of laws needs further development, as Lewis would agree, but he argues that the criteria of simplicity and strength are "safely noncontingent," and so unproblematic from a Humean point of view, and that the rest is just regularity. I am skeptical about both the clarity and the noncontingency of the criteria of simplicity and strength, but this is not my main worry, so I will grant the Humean seal of approval that Lewis places on his account of laws. The next step, and our main concern here, is the specification of the respects of similarity that are relevant to the interpretation of counterfactuals—a specification that makes appeal to the previously explained notion of law. From there he goes on to the star of Hume's family of problematic concepts, the concept of causation, which is analyzed in terms of causal dependence, which is analyzed in terms of counterfactuals.

Here are the proposed criteria of comparative similarity, given as an ordered list:

(1) It is of the first importance to avoid big, widespread, diverse violations of law.
(2) It is of the second importance to maximize the spatio-temporal region throughout which perfect match of particular fact prevails.
(3) It is of the third importance to avoid even small, localized, simple violations of law.
(4) It is of little or no importance to secure approximate similarity of particular fact, even in matters that concern us greatly.[7]

Lewis grants that there remains "plenty of unresolved vagueness" in this list of priorities, but he claims that it says enough to avoid the counterexamples that depend on the mistaken assumption that the analysis is based on an intuitive notion of overall similarity. My concern here is not whether counterexamples are avoided, but whether the specification appeals only to features of a possible world that satisfy the Humean strictures. As I said, I will set aside worries about the "best system" account of laws. The question I want to focus on is whether the defender of Humean supervenience can give a satisfactory account of particular fact, and of perfect match of particular fact. These facts, according to the specification of Humean supervenience, are constituted by the instantiation of intrinsic qualities by point-sized momentary objects, or perhaps by space-time points themselves. But what are

[5] Lewis, 1986a, *ix-x*. [6] Ibid., *xi*. [7] Lewis 1979b, 47–8.

these qualities? For Hume, they would be simple ideas that are copies of impressions, but Lewis's version of the thesis disclaims any commitment to the phenomenalist aspects of Hume's metaphysics—"It is no part of the thesis that these local matters are mental."[8] On Lewis's metaphysical picture, the local qualities exemplified in the actual world are the fundamental properties that physical theory tells us about; more generally, in an arbitrary possible world, they are the *perfectly natural* properties that are instantiated in that world. It seems, at least prima facie, that properties of this kind (mass, charge, spin, the colors and flavors of quarks), unlike the phenomenal properties that Hume might have taken as fundamental, are dispositional properties whose causal powers are essential to them. The defender of Humean supervenience must claim, as Lewis does, that this prima facie appearance is mistaken. The world can be described in terms of fundamental properties that are metaphysically independent of the laws governing the things that exemplify them, and of the way that things with those properties behave. I will consider in the next section Lewis's general account of natural properties, and his argument that they are independent of their causal powers, but first let me conclude with a general comment about what is required for a satisfactory defense of a reductionist thesis.

There is a temptation to describe any dispute between a reductionist and an anti-reductionist (for example, between a materialist who argues that mental properties are reducible to physical properties and a dualist who denies this) in this way: the question is whether a description of all the facts of a certain kind (for example the physical facts) would be a *complete* description, or whether there are features of the world over and above those facts. The reductionist who is not an eliminativist about a range of facts that appear not to be of the favored kind (the commonsensical fellow who doesn't deny the existence of the facts) is obliged to explain how the apparent further facts (for example, mental facts) are reducible to, or supervenient on, facts at the primary level. This characterization of the issue about a reductionist thesis may be appropriate when the question is about materialism, since both materialists and dualists may agree about what the physical facts are (those that can be described in terms of the resources of physics—perhaps some extension or revision of present-day physics). There may be some issues about exactly what the unproblematically physical facts are, but the disputes between materialists about the mind and their opponents do not depend in any obvious way on how the level of physical fact is understood. But this way of characterizing a dispute between a reductionist and an anti-reductionist requires the presupposition that a category of fact to which all facts either are or are not reducible has been identified. In the case of Lewis's more abstract and general reductionist program, this presupposition cannot be taken for granted. One may object to the thesis of Humean supervenience, not because one thinks that there are causal facts that go beyond the facts about the local qualities of point-sized things, but because one thinks that we have no notion of local quality that does not already have a causal dimension. On one influential account, properties in general are causal powers.[9] Such an account will imply that one cannot even begin to describe the basic facts that characterize the world, or any possible world, without making use of

[8] Lewis 1986a, *ix*. [9] See Shoemaker 1980.

notions that the Humean requires be reduced to something more basic. Lewis recognized the need to respond to this kind of objection, and his metaphysical theory addresses it explicitly. I will consider some of the consequences of the theory, and then close by considering how the problem of counterfactuals, and the more general problem of natural necessity, looks if one rejects the metaphysical picture, and the project of Humean reduction.

5. Natural Properties

Properties, for Lewis, are just sets. They are sets that may include merely possible things, which, according to Lewis's modal realism, are as real as any actual thing, so the identification of properties with sets is not stuck with the untenable consequence that properties that are coextensive *in the actual world* are thereby identical. Every set of possible individuals is a property, in the abundant sense of "property," but some properties are more natural than others, and some are perfectly natural. What is it for a property to be perfectly natural? Naturalness is of course itself a property—a property of properties—and so it is also a set—a set of sets. The property of being natural is one of the primitives of the metaphysical theory, to be explained in terms of the philosophical work that it does. Natural properties are assumed to be intrinsic, and it is assumed that there are enough natural properties so that one could give a complete specification of a possible world by specifying a spatio-temporal structure, locating a subset of the possible individuals in that structure, and saying what perfectly natural properties each of the individuals has (which natural sets it is a member of). Naturalness is a primitive notion, but we can get some grip on it by the examples of properties that Lewis takes to be paradigm cases of perfectly natural properties: the properties that our most fundamental science identifies are the perfectly natural properties that are exemplified in our world. These properties will be exemplified in some other possible worlds as well, but there will also be natural properties that are alien to our world, which means that they are unexemplified there.

Might the natural properties be dispositional? Might their nomological roles be essential to them? Lewis has an argument that the answer should be no. In fact, he goes so far as to claim that there is a possible world in which very different natural properties (a quark color and a quark flavor, in his example) trade places. "The two possibilities are isomorphic, yet different."[10] Here is the argument against the hypothesis that the nomological role of a property is essential to it:

> Start with a world where the quark colours and flavours do figure in the laws that are supposed to be essential to them. By patching together duplicates of things from that world, we can presumably describe a world where those laws are broken; yet perfectly natural properties are intrinsic *ex officio*, and so they never can differ between duplicates. The principle of recombination seems to me very compelling indeed.[11]

Duplicates, for Lewis, are individuals that share all intrinsic properties. The principle of recombination is (roughly) the principle that if there is a possible world containing

[10] Lewis 1986b, 162. [11] Ibid., 163.

an individual *a* and a possible world containing an individual *b*, then there is a single possible world containing both a duplicate of *a* and a duplicate of *b*. The basic idea of the argument is that if the principle of recombination is true, then no intrinsic property can have its nomological role essentially.

As a response to a critic of Humean supervenience, this argument has little dialectical force, since the critic has no commitment to a principle of recombination, or to the *ex officio* requirement that the most fundamental properties of things are intrinsic. Further, the denier of Humean supervenience may question the assumption (implicit in the order of explanation in Lewis's reductionist project) that the notion of a law of nature is the most basic notion in the natural-necessity family—that causal powers and dispositional properties are best explained in terms of laws, which are a kind of global regularity. For the Humean, causation is not a local relation, but depends constitutively on patterns of events throughout history, and so the causal powers, capacities, and dispositions of a thing will also depend on global patterns of the behavior of that thing and others—patterns that are described by the laws. But the anti-Humean rejects this assumption, taking causation to be a local relation, and this allows that the causal powers of an object may be intrinsic to it. Laws are presumably generalizations about causal facts and causal powers, but the anti-Humean rejects the assumption that causal powers are derivative from the generalizations about them, and so rejects the assumption that his claim that a certain property is essentially causal or dispositional can be put as a claim about nomological role.

The upshot is that Lewis's argument that natural properties are independent of their causal or nomological role presupposes Lewis's Humean metaphysical theory, and is not an argument for it. This does not deprive the argument of interest; I think it is best understood as an argument that helps to bring out some of the consequences of the metaphysical picture. But if the consequences are implausible, the argument may help to motivate a rejection of the premises that have these consequences.

The argument brings out that the metaphysical theory underlying Lewis's Humean project seems to require a radical gulf between the essential nature of a fundamental property—its quiddity—and the ways that the property manifests itself in the world. Lewis is up front about this commitment: as we noted, he holds that there is a possible world like ours except that a certain quark flavor and a certain quark color trade places. That is, in this possible world, things with the color behave exactly the way things with the flavor behave in our world, and vice versa. The distribution of the colors is exactly like the distribution of flavors in our world, and vice versa. Let's look more closely at the conception of natural property that this seems to require.

This alleged possibility of interchange of properties calls to mind Goodman's notorious predicates, "grue" and "bleen": (an object is grue at any time t iff it is green at t and first examined before some fixed time (in the future), T, or else blue at t and first examined after T. "bleen" is defined in a similar way, with "green" and "blue" interchanged. (See Goodman 1983, ch. 3). Let me pursue the analogy. If the interchange Lewis considers is possible, then so is a world where the quark color and the flavor (just for definiteness, let's say they are *green* and *strange*) trade places within the world at a certain fixed time. So, let our gruish (or perhaps grangish) world be one where at midnight GMT, January 1, 2000, these two very different fundamental properties trade places. *Green* and *strange* each take on the nomological role of the other, and all the green quarks suddenly become strange, and the strange quarks become green. (I don't understand any of this, which probably helps, but if these natural properties are as detached from their roles as Lewis claims, then this supposition should be unproblematic.) Perhaps this interchange actually happened in our world—who could tell? Maybe there was a catastrophic millennial bug after all, but no one noticed. If this change in fact happened, then for the past decade and more physicists have been systematically confusing greenness and strangeness. The true laws, which are those of the best systematization of the true natural properties, will be laws that mark the dramatic switch at the start of 2000 AD, but we (or the people in the world I am describing) will never discover these laws. Now remember that properties are nothing but sets—it is just that only some sets are *natural* properties. Consider not the property *green* itself, but instead the nomological role of *green* (now assuming that the dramatic switch did not in fact take place, so that the actual nomological role of the quark color is much as physicists take it to be). The nomological role property has the same extension in our world as the property *green*, but in other possible worlds, such as the grangish world I described, they have different extensions. Still, the nomological role property is a perfectly good property (in the abundant sense of property). Why is it less natural? One might think that a small change in Lewis's metaphysical theory could allow it to accommodate the thesis that natural properties are causal powers. Just say that a slightly different set of properties is the set of natural ones: the nomological role properties. But if we took the natural properties to be defined by nomological role, then even granting the primacy of laws in the order of analysis, and the best systems account of laws, this would compromise the Humean reduction, since the best system of laws and the choice of natural properties in terms of which the worlds are described would be interdependent. And as Lewis's argument shows, it would also cast doubt on a principle of recombination, or on the assumption that natural properties must be intrinsic.

The Humean project (not Lewis's but Hume's, and that of the twentieth-century empiricists, such a Goodman) was originally motivated by empiricist epistemological principles. Empiricists in this tradition took regularities to be less problematic than unobservable causal processes and underlying "occult" powers because generalizations about the phenomena, even if they faced a problem of induction, were on the surface, and were at least falsifiable. If I have interpreted Lewis correctly, then his version of the project leaves that motivation behind, and in fact requires a radical disconnect between the metaphysical account of reality and the epistemological account of how we know about it. If the natural properties that are the referents of our terms are as isolated from the laws by which those properties manifest themselves as Lewis's metaphysical picture seems to imply, then it is hard to understand how we ever know what the laws are, or how the properties of things connect with how the things with those properties behave. Even if a satisfactory epistemology could be combined with Lewis's metaphysical picture, it is no longer obvious what motivates the Humean strictures. Why, if not for the empiricist's reasons, is the family of concepts involving some kind of natural necessity problematic? I don't want to suggest that these concepts are not problematic, but if we give up on the project of

Humean reduction, then we need to say more about what the problem is. I will conclude, in section 6, with a brief remark about this.

6. Conclusion

In motivating the analysis of counterfactuals in terms of what is true in "close" possible worlds, Lewis considers the following challenge: "It's the character of our world that makes some A-worlds be closer to it than others. So after all it is the character of our world that makes the counterfactual true—in which case, why bring the other worlds into the story at all?" The reply is that "it is only by bringing other worlds into the story that we can say in any concise way what character it takes to make the counterfactual true. The other worlds provide a frame of reference whereby we can characterize our world" (Lewis 1986b, 22). For the defender of Humean supervenience, the frame of reference allows us to describe the actual world more concisely, and at a certain level of generality, but for the anti-reductionist, the possible-worlds frame of reference is playing an indispensible role in forming the concepts by which we describe our world at the most fundamental level. To describe the world just is to locate it in a space of possibilities, and one cannot separate the task of describing the world from the task of characterizing the space of worlds and the way our world is related to others. The development of a theory for predicting and explaining the phenomena involves the formation of the concepts by which the phenomena are described, and the testing of a theory against the evidence is also the testing of the legitimacy of the concepts that are formed.

The constructive task that remains if one replaces the Humean reductionist project with this more holistic picture is to develop a substantive account of the way that reference to other possibilities, and to a structure of relations between possibilities, contributes to fixing the meaning and reference of the terms we use to describe and explain the phenomena. This can be expected to involve connecting the inductive principles by which we test hypotheses with the principles by which we form the concepts that are used to structure the space of possibilities that our hypotheses distinguish between. More generally, it can be expected to involve clarifying the relation between the objective modal concepts and analogous subjective or epistemic modal concepts (causal necessity and epistemic certainty, chance and credence, causal independence and epistemic irrelevance, subjunctive and indicative conditionals). The idea of making this kind of connection has its roots in Hume, who emphasized the central role of causation in inductive reasoning, and connected the problem of induction with the problem of causation. It also connects with Goodman's observation that the problem of dispositions and the problem of induction are two aspects of a common problem. The conceptual projection involved in the forming of dispositional concepts exploits the same principles as the inductive projection involved in confirming theories. The abstract semantics for conditionals of the kind that David Lewis and I proposed for conditionals offers a framework for generalizing this kind of connection.

13

Dispositions and Chance

1. Introduction

One of the central projects in the philosophy of science of the first half of the last century was to explain the vocabulary of scientific theory in terms of the resources of an observation language. Dispositional predicates—terms like "soluble," "flexible," "brittle"—were regarded as the simplest kind of theoretical term, and a lot of attention was given to their analysis.[1] The interest at this time in the problem of counterfactual conditionals was motivated by the thought that if we could give a satisfactory account of them, we would have the resources to explain dispositional predicates, and ultimately all of the theoretical vocabulary of science. In his classic paper published in 1946, Nelson Goodman gave a clear characterization of what a reductive analysis of counterfactuals would require, one that set a demanding standard, but after making some tentative proposals and shooting them down, he argued that no analysis of this kind could succeed. When he returned to the problem a few years later in a set of lectures, publishing as *Fact, Fiction and Forecast*, he suggested that perhaps the problem of dispositions "is really simpler than the problem of counterfactuals," and that we might begin with it. But he did not offer anything like an analysis that would meet the conditions that he required for a reductive explanation of counterfactuals. What he did instead was to connect the problem of dispositions to a familiar problem in epistemology:

> The problem of dispositions looks suspiciously like one of the philosopher's oldest friends and enemies: the problem of induction. Indeed, the two are but different aspects of the general problem of proceeding from a given set of cases to a wider set. The critical questions throughout are the same: when, how and why is such a transition legitimate?[2]

He then turned his attention to the epistemological issues, first distinguishing the classic problem of induction (the problem of justifying the general practice of making inductive inferences) from what he called "the constructive task of confirmation theory"—the problem of describing the practice that we judge to be legitimate, or more specifically the problem of distinguishing those generalizations that are confirmed by their instances from those that are not. He argued that the first problem should be dissolved, but that the second problem is more challenging that one might have thought, and requires a new approach. I agree with all of this, but the particular constructive project that he began in that book did not (I believe) prove fruitful since it was tied too closely to Goodman's nominalism and extensionalism. I think there is

[1] See, for example, Carnap 1936. [2] Goodman 1983, 58.

a better way to develop his insight about (as I would put it) the relation between the practice of inductive inference and methods of concept formation.

Here is my plan: In section 2, I will sketch the problem of dispositions, and Goodman's reasons for seeing it and the task of constructing a confirmation theory as two aspects of the same general problem. In section 3, I will put Goodman's project into a wider context, seeing it as a version of what I have called the projection strategy, a strategy that has its origin in Hume's account of the role of the notion of causation in inductive reasoning. I will contrast two ways of understanding the projection strategy: (1) as the mistake of attributing to the external world features that are really aspects of the mind, and (2) as a legitimate method of concept formation. I will argue that (2) is the right way to understand the strategy as applied to dispositions, as well as to other objective modal concepts. In section 4, I will look in a general way at the interaction of the task of articulating an account of our methods of confirmation with the task of characterizing dispositional concepts. In the remaining sections I will look in more detail at an account of one kind of dispositional property—propensity—where the conceptual connection between epistemic and objective concepts is explicit.

2. Dispositional Properties

A *dispositional* property is a property that is understood in terms of the way it manifests itself under certain conditions. So, a disposition of an object is conceptually related to two other properties that the object may have: a *manifestation* condition, and a *test* condition. A glass is fragile if it would shatter if dropped, flexible if it would bend if subjected to suitable pressure, soluble (in water) if it would dissolve if put in water, observable if it could be seen (by a suitable observer) under suitable lighting conditions.

It used to be assumed that the problem of dispositions could be reduced to the problem of counterfactual conditionals in the following way: Where D is the dispositional predicate (e.g., flexible), M the manifest predicate (e.g., flexes) and T the test predicate (e.g., is subjected to suitable pressure), then D can be defined as follows: Dx if and only if ($Tx \Box\!\!\rightarrow Mx$). (An object is flexible if and only if it would flex if subjected to suitable pressure.) It was later argued, in a classic paper by C. B. Martin,[3] that this simple analysis will not work, and this led to an extensive literature about what came to be called finkish dispositions. Martin's example was what he called an "electro-fink": this is a device that detects when a live wire is about to be touched by a conductor, and under that condition instantly causes the wire to become dead. The property of being live is a dispositional property of the wire, where the test condition is "being touched by a conductor," and the manifestation condition is a current flowing from the wire to the conductor. The live wire has the dispositional property, and it is still live when connected to the electro-fink. But it does not then have the counterfactual property, "being such that a current would flow from the wire to the conductor, if touched by the conductor."

[3] Martin 1994.

A potential gap between counterfactual predicates and dispositional predicates was anticipated by Goodman, who gave a simple example of an object that is inflammable even though it would not burn if heated, since it is in an oxygen-free environment. The general lesson that Goodman drew was that "to speak very loosely, the dispositional statement says something exclusively about the 'internal state' of *w*, while our original counterfactual says in addition something about the surrounding circumstances."[4] His observation was not presented as a counterexample to the conditional analysis as Martin's argument was. Rather, his suggestion was that there is some flexibility in the statement of the test condition. One needs to build into it all the external conditions that are necessary for the manifestation condition to be realized in order to retain "full convertability between dispositional and counterfactual statements." But this means that we will be "forced back to some such fainthearted counterfactual" as

> If all conditions had been propitious and *w* had been heated enough, it would have burned.[5]

Instead of trying to make sense of such fainthearted counterfactuals, Goodman turned his attention to dispositions themselves, beginning by developing the analogy between the move from manifest to dispositional predicates and the inductive move from observed to unobserved instances of a generalization.

> Now "flexes" and "fails to flex" are mutually exclusive, and together they exhaust the realm of things that are under suitable pressure; but neither applies to anything outside that realm. Thus from the fact that "flexes" does not apply to a thing, we cannot in general infer that "fails to flex" does apply. Within the realm of things under suitable pressure, however, the two predicates not only effect a dichotomy but coincide exactly with "flexible" and "inflexible." What the dispositional predicates do is, so to speak, to project this dichotomy to a wider... class of things, and a predicate like "flexible" may thus be regarded as an expansion or projection of a predicate like "flexes".[6]

What is the basis for this projection or extension of the application of the predicate from the narrow domain of thing subjected to the test condition to a wider domain? Goodman's answer is that one uses exactly the same principles that one uses to make predictions about the future, or more generally to extend one's judgment about a narrow domain of things observed to a wider domain that includes things as yet unobserved. What principles are those? It is here where Goodman turned his attention to the problem of induction. But the problem, he argues, is not the traditional problem. "What is commonly thought of as the Problem of Induction," he said, "has been solved, or dissolved."[7] The problem is not to justify the practice of making inductive inferences in the way we do, but to characterize that practice, and then to use the distinctions we make in engaging in that practice to explain the formation of new concepts we can use to describe the world that we are making inductive inferences about.

[4] Goodman 1983, 39–40. [5] Ibid., 39. [6] Ibid., 44. [7] Ibid., 59.

In his discussion of the problem of induction and the "constructive task of confirmation theory," Goodman took himself to be following Hume, who set the precedent for connecting the practice of inductive inference with the development of concepts for describing the world. Inductive practice, according to Hume, essentially involves the notion of causation:

> All reasonings concerning matters of fact seem to be founded on the relation of *Cause and Effect*. By means of that relation alone can we go beyond the evidence of our memory and senses.[8]

But where do the ideas of cause and effect come from? All ideas are supposed to be copies of impressions, but we have no impression of causation. Here is Goodman's paraphrase of Hume's answer to this question: "When an event of one kind frequently follows upon an event of another kind in experience, a habit is formed that leads the mind, when confronted with a new event of the first kind to pass to the idea of an event of the second kind. The idea of necessary connection arises from the felt impulse of the mind." This is an explanation of an error, since the unconfused idea of necessary connection is the idea of a relation of ideas, and not of a relation between matters of fact. "Upon the whole, necessity is something, that exists in the mind, not in objects, nor is it possible for us ever to form the most distant idea of it, consider'd as a quality in bodies."[9] But Hume's answer to the question is also an explanation that implicitly endorses the notion of causation that arises from the error, since the explanation is a causal one. This is a reflection of the familiar tension in Hume's philosophy between his skepticism and his naturalism.

3. Two Faces of the Projection Strategy

The general Humean idea is that the concept of necessary connection between distinct events is the result of a projection of a feature of the mind onto the world, and one might appeal to this kind of projection more generally to explain the relation between subjective and objective modal concepts: epistemic relevance and causal dependence, credence and chance, indicative and subjunctive conditionals. But what is the metaphysical status of the statements that are made with concepts that result from this kind of projection? Here is the austere empiricist answer to the question—the Humean metaphysical picture, as described by Simon Blackburn:

> In this picture the world—that which makes proper judgements true or false—impinges on the human mind. This, in turn, has various reactions: we form habits of judgement and attitudes, and modify our theories, and perhaps do other things. But then—and this is the crucial mechanism—the mind can express such a reaction by "spreading itself on the world." That is, we regard the world as richer or fuller through possessing properties and things that are in fact mere projections of the mind's own reactions: there is no reason for the world to contain a fact corresponding to any given projection. So, the world, on such a metaphysic, might be much thinner than common sense supposes it.[10]

[8] Hume 1748, Section IV. [9] Hume 1896, 104–105. [10] Blackburn 1980, 75.

A. J. Ayer described a similarly austere picture of what there is a fact of the matter about:

I propose, then, to look upon the world as consisting of a bedrock of fact, and the only statements which I shall regard as being strictly factual will be those that are limited in their content to supplying true or false descriptions of this world, together with such statements as are obtainable from them by quantification or by the use of extensional operators. All other empirical statements, or at least all those that function at a higher level, will be construed as relating to the arrangement, or the explanation, of what are taken to be the primary facts.[11]

Ayer makes clear that what he has in mind by the primary system of fact is a system in which things are identified and described only by their phenomenal properties. He recognizes that ordinary language describes the world in causally loaded terms, but says that he wants "to make a distinction which ordinary language blurs." The terms used for describing the primary system of fact "are not be understood as carrying any logical implications about their powers."[12] Statements that concern the arrangement and explanation of the primary facts are not really statements of fact at all, according to Ayer:

The upshot of this discussion is that in a certain sense causes are what we choose them to be. We do not decide what facts habitually go together, but we do decide what combinations are to be imaginatively projected. The despised savages who beat gongs at solar eclipses to summon back the sun are not making any factual error.... They see what is going on as well as we do; it is just that we have a different and, we think, better idea of the way the world works.[13]

This starkly anti-realist (and I would say bizarre) view about the level of theory does conclude with an endorsement of "our" way of arranging and explaining the facts, but the judgment that it is better to arrange the facts as we do is given a noncognitivist interpretation.

I think there is an instability in Ayer's conception of "the bedrock of fact". On the one hand, this constitutes all that there is a fact of the matter about. But on the other hand, Ayer says: "It is only at some level of theory that we can form any picture of an objective world."[14] This seems to imply that it is not a matter of objective fact that we live in an objective world. I would prefer to say that we can agree that a conception of an objective world is formed only with the development of theory, while also taking that level of theory to make a factual claim about what the world is like.

Blackburn expresses sympathy for the view that the objective world "might be much thinner than common sense supposes it," and he does defend an expressivist, or to use his term, quasi-realist account of normative judgments, but he does not defend the particular anti-realist thesis that Ayer's remarks seem to imply. He recognizes that there is a continuum of views about how widely what he calls the Humean mechanism should be applied, and his argument is only that there is more to be said for the application of this mechanism than is often supposed. I agree with Blackburn's expressivism about norms, including epistemic norms, and with the quasi-realist formulation of that view that helps frame and sharpen the debate about

[11] Ayer 1972, 115. [12] Ibid., 115. [13] Ibid., 138–9. [14] Ibid., 114.

what there is a fact of the matter about.[15] But I will argue that even if our theoretical vocabulary about causal powers, dependence and independence, capacities, potentialities, and dispositions has its origin in a projection of habits of mind onto the world, one can take that kind of projection as a legitimate form of concept formation with products that can be used to formulate hypotheses that are tested, using inductive rules that we endorse, and accepted as factually true. There is a kind of circle here—an unavoidable feature of the kind of dissolution of the old problem of induction that Goodman and others defend. We form a picture of the world that includes ourselves as things we find in it, things that have the capacity to receive and use information about the world. Part of our theory is an explanation of why the methods we use work to tell us what that world is like. (This is a way to understand Hume's skeptical solution to his skeptical problem about induction.[16]) The inductive methods are developed and refined together with the formation of concepts for stating hypotheses about the world that we are learning about with those methods. In the next section, I will look in a general way at the interaction of the project of articulating reasonable inductive practice with the project of forming the concepts for describing the world, particularly concepts of dispositional properties.

4. Inductive Practice and Dispositions

Confirmation, as Carnap and Quine taught us, is holistic. From an a priori perspective, anything might be evidence for anything else, since relations of evidential relevance depend on what information we already have. If you happen to know that some disjunctive statement, P or Q, is true without knowing which, then even if the two propositions have nothing to do with each other, learning that one is false will be decisive evidence that the other is true. It might be that for Zack, learning that P would be decisive evidence for Q, while for Jack, learning P would be decisive evidence for not-Q, and so long as P is in fact false, it may be that neither Jack nor Zack is making any epistemic mistake in being disposed to make these inferences.[17] But even if we cannot generalize in an unqualified way about what is evidence for what, our attempt to give a systematic theory of confirmation will be in part a search for stable features of the world about which qualified generalizations about relevance and irrelevance can be made. Suppose we can identify hypotheses that satisfy the following condition: if they are true, then we can be sure that certain relations of evidential relevance will hold (certain further information would be evidence for other further facts.) Dispositional properties can be expected to be involved in hypotheses that play such a role.

[15] See Chapter 7 of this book. [16] cf. Chapter 1 of Stalnaker 2008.

[17] I am alluding here to Allan Gibbard's notorious Mississippi Riverboat example, which he uses to argue against a propositional analysis of indicative conditionals. Sly Pete is in a poker game, and is either to call or to fold. Jack has seen both hands, and knows that that Pete has the worse hand, and so if he called he lost, while Zack knows that Pete knows the contents of both hands, and so will call only if he has the winning hand. So, it must be that if he called, he won. I discuss the example in a number of places, including in Stalnaker 1984, ch. 6, and in Stalnaker 2011c, reprinted as chapter 11 of this book.

Goodman was a nominalist who was happy enough to talk of predicates and their extensions, but not about properties. This, however, is a hang-up that we can separate from his projection strategy for explaining dispositions. One who is friendlier to properties might put Goodman's point this way: We find a distinction (things shatter, or they don't) that is manifested only under a certain test condition (when they are dropped). We hypothesize that there is a property that coincides with the display property under the test condition (at least under normal conditions), but that things have, or lack, *independently* of whether they satisfy the test condition. The test and manifestation conditions are used to fix the reference of this hypothesized property.

The hypothesis that there is a property that is manifested in a certain way under certain test conditions, but that things may have independently of whether they are subjected to test is an empirical hypothesis that may prove to be false. For example, some have thought that there is dispositional property of "hot-hands"—roughly, the property that a basketball player may have of being more likely to make a shot after having just made one. The opposing view is that normal random variation in an individual player's performance gives rise to the illusion that there is such a property. One would have evidence for the presence of such a dispositional property if one found that a successful shot was positively relevant to the prediction of success on that player's next shot. The evidential situation is delicate, but the hypothesis that there is such a property, and that certain players have it is subject to empirical test. In many cases, where there is evidence that the hypothesized property correlates with other properties, the evidence for it can be very robust.

This kind of concept formation begins with relatively superficial dispositional properties such as flexibility and fragility, where the test conditions and the manifestations conditions are somewhat vague, but the hypothesis that there is such a property presupposes that there will be a causal explanation for the fact that the things with the property behave as they do. Our theorizing moves from more superficial dispositions of this kind, to deeper ones such as solubility or electrical conductivity, where the manifestation conditions (dissolving, or carrying an electrical current) are themselves theoretical. It is common in discussions of dispositions to talk of a causal basis for a disposition, but it is a mistake to think that the hypothesis that a dispositional property has a certain causal base (for example, that an object is fragile because it has a certain molecular structure) is an explanation for a disposition in terms of something categorical, where "categorical" contrasts with "dispositional." The causal base is just a deeper and more theoretical disposition, though perhaps one that may be manifested in multiple ways under different test conditions. It seems reasonable to suppose that even when we reach properties defined in terms the most fundamental features of the most fundamental physical entities, our properties remain dispositional in the sense that the way they are manifested under various conditions is essential to them.[18]

[18] Hugh Mellor argues for irreducible dispositional properties in Mellor 1974. See also Sydney Shoemaker's defense of the thesis that properties generally are causal powers in Shoemaker 1980. A contrary view is Lewis's conception of natural properties, defended in Lewis 1983b. On this view the relation

The general strategy is to try to find stable features of the world that are located in particular parts of it and that are conceptually connected to stable patterns of inductive inference. The intuitive picture is something like this: If we look just at the surface of things, we find a seemingly chaotic sequence of events as things interact with each other. The theorist hypothesizes that the things that are interacting have certain underlying properties that explain the interaction—properties that are more stable and independent than the overt behavior of the interacting things would suggest. When two or more things interact, the behavior that is manifested of course depends on the underlying dispositional properties of all of them, and the test conditions for the dispositions of one thing may make reference to the dispositional properties of the things that it may interact with. To use a familiar example, the key is disposed to open a certain kind of lock. What kind? The kind that is disposed to be opened by the key. It is obvious that there is no analysis of the dispositional properties here, but there are nontrivial hypotheses implicit in this characterization that help to define the explanatory project of determining the character of the relevant properties of the separate components of the system.

Dispositions come and go, and a thing may be disposed to acquire or lose a disposition. The manifestation condition for one disposition may be the acquisition of another disposition. The banana is not fragile, but it is disposed to become fragile, and so to shatter when dropped, if super-cooled. The ice sculpture is disposed to shatter when dropped, but also to melt when heated, and so to lose its fragility when heated. The finkishness of dispositions in certain contrived situations—the phenomenon that demonstrates the inadequacy of the simple conditional analysis—is just one kind of example of the complex interaction of dispositional properties. The general pattern of a finkish disposition is something like this: Property D1 is the property of being disposed to manifest M in condition T, and D2 is the property of being disposed to lose D1 in condition T. D1 is a finkish disposition of an object that also has dispositional property D2. (In the standard examples, D2 is an extrinsic dispositional property, but it could be a separate, but still intrinsic disposition property.) Finkishness, along with other patterns of interaction of dispositions, show that the stability and independence that dispositional hypotheses are aiming at is relative and defeasible: when a disposition D1 is finkish, the test condition and the property D1 are not, because of D2, counterfactually independent. But a notion of counterfactual independence still plays an essential role in the characterization of dispositions. It is essential to an example of a finkish disposition that the two dispositional properties, D1 and D2, be independent at least in the sense that it makes sense to suppose, counterfactually, that the object has the one, but not the other. If you try to construct an example of a finkish disposition where D1 and D2 are realized in the same underlying property, you will fail.

A notion of causal independence is essential to the specification of a dispositional property, and relations of causal dependence and independence will be essential more generally for explaining why certain information is evidentially relevant or

between natural properties and causal powers is contingent. I criticize Lewis's notion of natural property in Stalnaker 2015, reprinted as Chapter 12 of this book.

irrelevant. The projectivist picture suggests that causal concepts are a kind of projection of evidential relations, but hypotheses about causal relations will still be subject to empirical confirmation, and so can be interpreted realistically. The connection between relevance and dependence will always be qualified and defeasible. One cannot infer, for example, from the fact that two possible events are causally independent that the occurrence of one is therefore evidentially irrelevant to the occurrence of the other, since the two events might have a common cause, or might be known to be similar in their causal properties. But it is reasonable to assume that a piece of information can be relevant to a hypothesis only if there is a causal explanation for the relevance. Our general account of knowledge will be a causal account—we know about the world by causally interacting with it—so it is inevitable that there should be conceptual connections between the rules we apply in acquiring knowledge and the concepts with which we describe the world we know about.

In the rest of this chapter I will look in more detail at an account of one kind of dispositional property where the conceptual connection between evidential and causal notions suggested by the projection strategy is explicit.

5. Propensities and Objective Chance

It is controversial whether there is such a thing as objective chance. Hume certainly denied it, as did some of the developers of the Bayesian account of degrees of belief such as Frank Ramsey and Bruno de Finetti. But it is agreed that there is an appearance of objective chance that plays a role in scientific reasoning, and that needs explanation. The projection strategy is natural here: the idea is that the notion of objective chance arises from a projection of rational credence onto the world. But I have argued that this strategy is compatible with a realist interpretation of the application of the concepts that result from this kind of projection.

Chance, or its appearance, is a feature of situations involving the interaction of objects. It is *propositions* that have, or appear to have, a chance of being true, but following the general pattern of the introduction of dispositional concepts, it will be features of the objects involved in the interaction that explain why events have the chances they seem to have. A *propensity* is a disposition, and on the account of propensities I will explore, the display of the disposition will be explained in terms of the degree of belief that it justifies. Here is Hugh Mellor's characterization of propensity:

> The relation between the propensity and personalist theories [of probability] is this: According to the latter the making of a probability statement expresses the speaker's "partial belief" in whatever he ascribes probability to, say that a coin *a* will land heads when tossed. Knowledge of the coin's propensity on the present theory is what in suitable circumstances makes reasonable the having of some particular partial belief in the outcome of the toss. The chance of the coin falling heads when tossed is then the measure of that reasonable partial belief.[19]

[19] Mellor 1971, 2.

The test condition for the propensity of the coin is the tossing of it in the appropriate chance set up, but the "display" is not the particular outcome of the toss. Rather, it is defined in terms of what it is reasonable to believe about the outcome. The coin has a propensity of .5 to land heads if it has a property that, if one knew about it, would license a degree of belief .5 that the coin would land heads if it were flipped.

Consider these two generalizations about chance that seem to be plausible, and in need of explanation:

(1) When the coin has (and retains) an objective chance of *r* of landing heads (in a certain coin-flipping set-up) then the limit of the frequency of heads in an infinite sequence of flips is *r*.
(2) If an agent knows that a coin has an objective chance of *r* of landing heads, then (before it is flipped) he or she should have a degree of belief *r* in the proposition that the coin will land heads.

Mellor's propensity theory takes (2) (a simple version of what David Lewis dubbed "the Principal Principle") to state a constitutive conceptual connection between objective probability and rational degree of belief. The more traditional approach to objective chance (a theoretical version of a relative frequency interpretation) is to take (1) to state a constitutive conceptual connection. On this approach, propensity should be explained as a disposition to produce a sequence with a certain relative frequency, as suggested in the following quotation from a probability textbook:

> It seems natural to postulate the existence of a number *P* which can be conceived as a mathematical idealization of the frequency ratio *f/n*, in the same way as the hypothetical true value of some physical constant constitutes a mathematical idealization of our empirical measurement. . . .
>
> *This number P will, by definition, be called the mathematical probability of the event A in connection with the random experiment E.*[20]

Mellor's analysis has the burden of explaining, given (2), why (1) should be true, while the more traditional theoretical frequency view has the burden of explaining, given (1), why (2) should be true. I don't see how to explain (2) in terms of (1), but an explanation of (1) in terms of (2) is straightforward: a rational agent who knows or assumes that the coin has a certain propensity and retains it will (according to (2)) have credences in the result of the different flips in a sequence that are probabilistically independent, so that learning the results of earlier flips will not influence her rational degree of belief in later flips. This implies that the rational degree of belief, on the condition that the chance of heads is *r*, in the proposition that a sequence of *m* flips will approximate a relative frequency *r* will approach one as the sequence increases.

David Lewis's Humean metaphysics requires that facts about chances be supervenient on global patterns of actual sequences of events, and so seems closer to a version of a theory of objective probability based on a constitutive connection between chance and frequency, but Lewis also argues that his Principal Principle,

[20] Cramér 1955, 26.

on some way of making it precise, should be a fundamental epistemological principle governing any reasonable account of chance.

> The distinctive thing about chances is their place in the "Principal Principle," which compellingly demands that we conform our credences about outcomes to our credences about their chances. Roughly, he who is certain the coin is fair must give equal credence to heads and tails.[21]

Lewis does not offer any explanation of how the relevant patterns of particular fact should be a basis for accepting the compelling epistemological principle; he expresses only a tentative intuition that such an explanation should be possible.

> I can see, dimly, how it might be rational to conform my credences about outcomes to my credences about history, symmetries, and frequencies. I haven't the faintest notion how it might be rational to conform my credences about outcomes to my credences about some mysterious unHumean magnitude. Don't try to take the mystery away by saying that this unHumean magnitude is none other than *chance*! I say that I haven't the faintest notion of how an unHumean magnitude can possibly do what it must do to deserve the name—namely, fit into the principle about rationality of credences—so don't just stipulate that it bears that name.[22]

I confess that my intuitive reaction is exactly the opposite of Lewis's. I think I see how an unHumean account of chance such as Mellor's can justify the Principal Principle, but I have no idea how a proponent of the thesis of Humean supervenience might explain a principle that seems to state a tight conceptual connection of this kind between a theoretical hypothesis about general patterns of events and an epistemological principle about what it is rational to believe about certain particular events. Furthermore, I think one can provide more than an intuitive hunch about how the justification should go on the unHumean analysis. Here is an attempt to explain it:

Mellor's propensity analysis, as I interpret it, uses the connection to rational credence to *fix the reference* of a property. As discussed above about dispositions in general, reference-fixing involves both an empirical hypothesis and a stipulation. The idea is that one *hypothesizes* that there is a stable property (a propensity with measure *r*) such that if one knows that an object (for example, a coin) has that property, then it is rational to have credence *r* in the proposition that a certain result will occur (that the coin will land heads), conditional on the object being subjected to a certain test condition (the coin being flipped). Then one *stipulates* that a certain term ("propensity of *r*") shall refer to that property, should there be such a property.[23] The stipulation, together with the probabilistic laws that govern credence allows one to test the hypothesis that a certain object has this hypothesized theoretical property. Specifically, the stipulation determines credence values (conditional on the hypothesis that the object has the property) for propositions about observable events (sequences of results of repetitions of the experiment). Then by Bayes' theorem, this, together with a prior credence in the hypothesis, and

[21] Lewis 1986b, *xv*. [22] Ibid., xv–xvi.

[23] cf. Ned Hall's discussion of "the skirmish over non-reductionism" in Hall 2004, 106–12.

a prior credence in the observable events, yields a credence value for the hypothesis conditional on the evidence. So long as the prior probability of the hypothesis is not negligibly small, frequency information will eventually provide very strong evidence for or against the hypothesis.

One might begin with a *general* hypothesis that the object has a stable propensity, without assuming any particular value, and then let the experimental results provide evidence for a particular value. Evidence against the general hypothesis might consist of a persistent pattern in the sequence of results of the experiment that supports some alternative hypothesis that allows for more accurate predictions of the outcomes than those provided by any propensity hypothesis. Hilary Putnam once gave a fanciful example of evidence of this kind.[24] Suppose, in a finite sequence of flips of a coin, one observed that all of the prime-numbered flips landed heads, while the others all landed tails. If the sequence were short, this would be judged an amusing coincidence, but if it persisted long enough, one would eventually reach the conclusion that the sequence must somehow be rigged. The prior probability (credence) of the hypothesis that the sequence was rigged to produce this result would be extremely small (or more likely, the hypothesis will not even be considered), but if the pattern were noticed, the hypothesis would become salient, and then increasingly probable on the increasing evidence.

6. Lewis's Principal Principle

Lewis formulates his Principal Principle as a constraint on a "reasonable *initial* credence function," C. The principle says that C(A/XE) = *x*, where A is any proposition, X is the proposition that the chance of A at time *t* is *x*, and E is any evidence that is *admissible* at time *t*. Lewis makes clear that what he means by a reasonable initial credence function is an a priori ur-prior that determines what our credences should be in any evidential situation we might find ourselves in. A reasonable initial credence function Lewis says, must be *regular*, which means that it should assign a positive probability to all metaphysically possible propositions. The initial function must be regular since "one who started with an irregular credence function (and who then learned from experience by conditionalizing) would stubbornly refuse to believe some propositions no matter what the evidence in their favor."[25] The presupposition of this last claim is that one's entire inductive method should be modeled by a probability function that is chosen independently of any empirical evidence, and that determines what one's credences should be in any conceivable situation. This is a presupposition that does not fit well with the process of developing and refining methods of confirmation in interaction with a process of developing and refining a theoretical picture of what the world is like,[26] and I think it should be rejected. But we don't need the notion of an ur-prior to state a defensible version of the Principal Principle, one that is a simple constraint on the actual ("posterior") credence function that it is reasonable to have at a given time.

[24] Putnam 1963, 765. Putnam's story did not involve coin flips, but the idea is the same.

[25] Lewis 1980b, 88.

[26] I have criticized the notion of an ur-prior in several of the chapters of this book. See Chapters 5 and 7.

There has been considerable discussion in the literature on Lewis's Principal Principle about what it is for evidence to be *admissible* at a certain time, but I will argue that we don't need any such restriction in a defensible version of the principle that is motivated by a propensity account such as Mellor's. The simple principle is this: $Cr_t(\phi/ch_t(\phi) = r) = r$. '$Cr_t$' is credence at t, '$ch_t$' is chance at t, ϕ is any proposition and r is any real number in the [1,0] interval. We do not need an evidence parameter in our principle, since we are starting with a credence function that has already taken account of the evidence that the agent has at the time, and any evidence the agent actually has will be admissible in the sense Lewis has in mind. (I will be assuming that our agent does not have access to oracles, crystal balls, time-traveling informants, or other backward-causing sources of information.)

The rationale for this constraint on credence (according to the propensity theory) is that it is a presupposition of the chance proposition, $ch_t(\phi) = r$, that there is no information that could be available to the agent at time t that would give her reason to have a degree of belief difference from the chance value r. If there were such information, then the relevant objects would not have the propensity properties that make the chance hypothesis true. The condition that there is no information of a certain kind that could be available is, at least implicitly, a causal hypothesis, since our conception of knowledge is a causal conception: a fact can be known only if the agent is in an epistemic state that is causally sensitive to a state of the world that entails that fact.

7. Exchangeability and Objective Chance

There is a famous technical result, published in the 1930s by Bruno de Finetti, that provided support for the projection strategy for explaining objective chance.[27] The rough idea is this: Start with an algebra where the basic "events" (propositions) are given by an infinite sequence of observable outcomes of a repeated experiment (say the results of the flips of a coin). Assume that the events are *exchangeable*, relative to a credence function defined on this algebra. What this means is that the probability of any proposition definable in the algebra (any truth-function of the basic events) will be equal to the probability of any proposition that is a permutation of the basic events.[28] The result is that any credence function meeting these conditions can be represented as a mixture of probability functions, each of which meets a condition that is stronger than exchangeability: the condition that the events are stochastically independent. The projectivist interpretation of the result is this: the probability functions in the mixture represent hypotheses about *apparent* objective probabilities. One could make the projection explicit by refining the algebra, adding theoretical propositions stating what the objective chances are of the basic observable events, and the result shows that the extension of the probability function to the propositions

[27] de Finetti 1964.

[28] A *permutation* of a set of events (propositions) is a 1–1 function of the set onto itself. A permutation of the basic events (propositions about the results of individual coin flips in the sequence) generalizes to a function that maps any truth function of propositions in the given set onto truth functions of the propositions that result from the permutation.

in this finer-grained algebra would be determined by the original probability function. The credences for those theoretical hypotheses would receive conditional credence, relative to purely observational evidence propositions, and the result would also show that enough of such evidence will bring the credences in the theoretical hypotheses as close to one or zero as one likes.

De Finetti was a resolute subjectivist who held what Blackburn called a "thin" Humean metaphysical view. He regarded his theorem as an explanation for the appearance of objective probability, and a justification for concluding that the application of probability in statistical reasoning in science could do without a realistic interpretation of it. He also took the result to provide an explanation for the utility of subjective probability for finding out about the world, but the result itself cannot help to justify the use of statistical reasoning in inductive practice unless one makes the exchangeability assumption, and one might ask what explains why the agent should have, in certain situations, credences that satisfy that condition. One answer to this question is that if we make certain *qualitative* assumptions about the causal structure of the sequence of events, then the exchangeability hypothesis will be justified. Suppose we assume, for the particular sequence, that the basic events are alike (and so all receive the same prior probability), and that they are causally independent of each other. Suppose we also assume, more generally, that there must be a causal explanation for epistemic relevance. It is compatible with these assumption that information about the results of some flips will be evidentially relevant to the result of others, but the assumptions justify the requirement that an agent's *prior* beliefs about the results of any set of flips (say HTH for the first three flips) should be the same as her prior beliefs about any corresponding set (say H on the fourth and seventh flip, and T on the tenth), and that is what exchangeability requires.

The technical result does not depend on any explanation or justification for the fact that the basic credence function satisfies the exchangeability constraint, and on the thin Humean metaphysic, causal notions are themselves mere projections. But the realist asks why it is not enough for a realistic interpretation of the concepts that result from such projections that hypotheses involving them can be confirmed or disconfirmed by evidence that is unproblematic from the Humean point of view (in the coin flip example, by observation of the results of the flips). Humeans justify their projections pragmatically: it is fortunate that we have habits of mind that, when we project them onto the world, help us to form reliable beliefs about matters of fact. But by the most general and truistic principles of inductive reasoning, hypotheses that generate reliable predictions about the facts are hypotheses that we have some reason to believe are true.

In simple situation like the coin flip examples, with an object that has an unchanging propensity that can be manifested over and over, hypotheses about objective chance can be straightforwardly confirmed or disconfirmed, but as we noted in the general discussion of dispositions, properties come and go. An object may have dispositions to acquire or lose a disposition, and the manifestations of the dispositions of an object will depend on the dispositions of the other things it interacts with. All of this will of course be true for propensities, which can change over time, and can be influenced by events, including chance events. Assumptions about the causal structure of the situation in which chance events take place are all the more important in the more complex situations. Consider the classic example of a Pòlya urn: Suppose an urn initially contains just one black ball and one red ball, and that one of them is drawn at random, and then replaced by two balls of the same color as the one that was drawn. The procedure is repeated over and over, with the propensity of the urn to produce a red ball changing with each repetition. One of the interesting things about Pòyla urns is that they provide a model for a probability function on events that are exchangeable, but not stochastically independent, but the urn model is very different in its causal structure from the model with a sequence of coin flips with a fixed but unknown chance. The Pòlya urn models show that, while the exchangeability condition can be justified by an assumption of causal independence, one cannot *derive* a belief in causal independence of the basic events from the fact that the credence function satisfies exchangeability. De Finetti's result still applies to the Pòlya urn model, but the probability functions that define the mixture cannot be reasonably interpreted as objective chance functions about which the agent has credences. If the agent begins with a prior credence function on the observable events in the sequence that just happens to satisfy the exchangeability condition, and then bases her posterior credences solely on information about those events, her credences will evolve in exactly the same way, whether the true underlying story is the urn story or the coin flip story. The Humean might be tempted to take the upshot of this point to be that the background causal stories we tell ourselves should not be taken seriously, and perhaps this would be right if all of our empirical evidence were about simple sequences of events of this kind. But it is also true that if our access to information about the world were this limited, we would have no reason to begin with a credence function that satisfied the exchangeability condition, or any other condition that goes beyond coherence. In the more complex and diverse epistemic situations we find ourselves in, different hypothesis about causal structure will make a difference to the observations we expect to be in a position to make, and so can be confirmed or disconfirmed. And in the wider context, hypotheses about causal structure will play a central role in developing and motivating inductive principles as we develop together a theory about what the world is like and a system of rules for learning about it.

8. Realism about Chance and Causal Structure

The "thin" Humean metaphysic exemplified by de Finetti, Ayer, and Ramsey is anti-realist about the level of theory that results from the projection of our habits of mind or epistemic policies onto the world: theory goes beyond what there is a matter of fact about. David Lewis's Humean metaphysic contrasts with this: he begins with a level of fact with a similar structure, but defends a realist account of causal notions, including chance, by arguing that the level of theory is reducible to, or supervenient on, a Humean base. The realist picture I am defending contrasts with both of these philosophical views since it rejects a presupposition that they share: that there is an autonomous level of fact that conforms to the Humean strictures. On my realist picture, we develop resources for describing the world while we are also developing a theoretical account of our place in it that explains how we are able to know about it. Causal structure is built into the most basic descriptive concepts that result from this process, but there will also be, on this picture, conceptual connections between these descriptive resources and epistemic principles. I will look very briefly at three issues concerning chance that help to bring out the differences between these three alternative general views: first, metaphysical and epistemic undermining, second, compatibilism about chance, and third, skepticism based on chance.

Metaphysical and epistemic undermining

Lewis's defense of Humean supervenience faces a problem concerning chance that he regarded as the most serious challenge to this metaphysical project. Here is the rough idea of the problem of undermining: on Lewis's Humean story, laws of nature are just descriptive generalizations about global patterns of events. The system of laws of nature that is true in a given possible world is a set of exceptionless generalizations that "achieve an unexcelled combination of simplicity and strength."[29] For some possible worlds, the generalizations that meet this standard will include nondeterministic laws that say that certain events have a certain chance of happening, but leave open the question whether they happen or not. Worlds with chance laws will be worlds where stronger generalizations that are deterministic would buy their additional strength at too great a cost to simplicity, and so will count as accidental generalizations. So far so good, but the problem is that any chance laws that satisfy the strength-and-simplicity standard in a possible world x will, because they are nondeterministic, be compatible with a different global pattern of particular fact that holds in a possible world y that allows for a set of deterministic generalizations that is simple enough to be the laws of world y. The point is most easily illustrated with a toy model of a world x where the Humean facts consist simply of an infinite sequence of events like coin flips. Suppose the limit of the relative frequency of flips in world x is .5, but no *simple* function would generate the sequence, or specify it in more detail. There is no simple pattern in the sequence: for example, no simple function for picking subsequences would do a better job than .5 job of predicting the outcomes for the events in that subsequence. So, the generalization that best fits the standard for a law in world x is the chance law, with the chance of heads fixed at .5. But in world y, the coin displays a simple pattern: perhaps all heads, perhaps alternating heads and tails, perhaps heads on prime-numbered flips and tails on the others. Any of these outcomes is compatible with the chance law of world x, which allows any sequence, but in world y, the chance laws are false since the simple deterministic function that generates the sequence will be both simple and stronger.

This is a problem because one wants to say that there are two distinct y worlds: Because the chance law of x permits the y pattern to be manifested, there should be a world y_1, where the chance law of x holds, and the coin's regular behavior is an improbable coincidence, but there should also be a world y_2 where the coin's regular behavior is entailed by a true law. But since y_1 and y_2 exhibit exactly the same global pattern of events, to say this would be to give up Humean supervenience.

The unHumean propensity theory does not face the metaphysical problem of undermining. Since it takes the causal structure that explains why a certain pattern

[29] Lewis 1986b, xi.

obtains to be a real feature of the world, it can distinguish y_1 from y_2. But there is a related phenomenon of *epistemic* undermining. As we noted in discussing the Putnam example, a chance hypothesis can be defeated by robust evidence of a pattern that allows for predictions that are more accurate than those given by the chance hypothesis, even though the defeating evidence is compatible with the hypothesis ("defeated" here means not that the hypothesis is definitively proven false, but only that one has sufficient evidence to reject it). Timothy Williamson gives this example[30]: suppose two balls, one red and the other black, are put in a bag, and one is drawn at random, and then replaced. The experiment is repeated a thousand times, and a red ball is drawn each time. Alice takes herself to know, at the start, that the bag contains one red and one black ball, and that the chance of getting a red ball on a draw is .5. Long before the end is reached, these claims to knowledge are defeated by the long sequence of all red draws, but Williamson asked us to suppose that the set-up was in fact as described, and that the resulting sequence was just a coincidence. He concluded that Alice really did know, at the start, that the set-up was as described, but that she lost that knowledge at some point along the way. I would prefer to say that because of the abnormality of the data that emerged as the experiment went along, Alice did not know at the start what she reasonably took herself to know: she was in a Gettier case. But either way we can consider the case where Alice reflects, at the start, on the possibility of getting this evidence: "Suppose," she says to herself, "I draw the red ball a hundred or more times in a row. What would I think then?" Her answer, like ours, will be that she would then be quite certain (would take herself to know) that the set-up was not as she thought: it must be that both balls are red, or that some trick has ensured that the red ball is drawn each time.[31] So if E is the evidence of 100 red draws in a row, from the start, and X is the hypothesis that the chance of drawing a red ball on the 101st draw is .5, her conditional credence, at the start, Cr(X/E), will be 0, or close to it. I am assuming here that conditional credence may be defined even if the credence in the condition is 0, as will be true in this case, assuming that Alice initially took herself to know that the set-up was as described, and so that Cr(X) = 1.[32] But we want conditional credence to be defined even when prior credence in the condition is 0 since the reflective person can consider what she would believe if she learned that she didn't know what she took herself to know.

Alice can also ask herself, as she reflects before the first ball is drawn, "Suppose 100 red balls are drawn in a row. Even though I would then be certain that the chance hypothesis must be false, I can still ask what my *conditional* credence should *then* be, conditional on the chance hypothesis being true, that the 101st draw will be heads." The answer (I think all will agree) should be .5. That is, if Cr*

[30] Williamson 2000, 205.

[31] If your epistemic policies would be more cautious than this, choose a number larger than 100.

[32] Lewis allows this by allowing infinitesimal probability values. I would prefer a representation with a lexicographic probability system, but the elementary laws of probability where conditional probability is primitive, are the same in either case. The main difference in this more expressive formulation of probability theory is that the ratio definition of conditional probability is replaced by an axiom: Pr(AB/C) = Pr(A/C) × Pr(B/AC). Absolute probabilities are defined as conditional probabilities, conditional on a tautology.

is the envisioned credence function, after learning E (so that Cr*(E) = 1, and Cr*(X) = 0)), $Cr^*(H_{101}/X) = .5$, and that is what the propensity theory says.

If the regular pattern were universal, so that even an agent who was omniscient about the observable events would come to believe that the chance of a red draw was 1, the agent's true beliefs about future outcomes would not constitute knowledge, since knowledge requires that the agent's beliefs about the outcomes be sensitive to facts that causally explain them. We can assume that the chance hypothesis is still true in such a case, and that the agent is in a skeptical scenario.

Lewis seems to agree about epistemic undermining. He considers this example:

> Suppose that radioactive decay is chancy in the way we mostly believe it to be. Then for each unstable nucleus there is an expected lifetime, given by the constant chance of decay for a nucleus of that species. It might happen—there is some chance of it, infinitesimal but not zero—that each nucleus lasts for precisely its expected lifetime, no more and no less. Suppose that were so... If the matter were well investigated, doubtless the investigators would come to believe in a law of constant lifetimes. But they would be mistaken, fooled by a deceptive coincidence.[33]

I think this is exactly right, but I am not sure how Lewis can reconcile what he says about this example with Humean supervenience, since he also says that this regularity, which is accidental, given the chance law, "might well qualify to join the best system... Still it is not a law." It is right that the regularity cannot be *both* a deterministic law and a low-probability coincidence, but according to Humean supervenience, the global pattern of particular events is supposed to determine which of them it is. This implies that nothing can be a regularity that would "qualify to join the best system" for a certain global pattern of events, while being "not a law" in a world given by that global pattern.

Compatibilism about chance

The second issue is whether chance is compatible with determinism. Lewis takes the hard line that the only chance that one should be a realist about is chance that derives from indeterminacy in fundamental physics, and if physics turned out to be deterministic, there would be no chances other than zero and one. He acknowledges (with some disdain) that one can have a theory of what he calls "counterfeit chance" which is "a relative affair, and apt to go indeterminate, hence quite unlike genuine chance."[34] But I think he exaggerates the difference between so-called counterfeit chance and the real thing. And it is not that that we need an account of chance that goes beyond its role in quantum theory simply to "serve the needs of determinists." We need it to explain the pervasive role of objective probability in science, and more generally in statistical reasoning in domains of inquiry far removed from fundamental physics, for example in epidemiological studies, or in the interpretation of polling data in the prediction of elections.

[33] Postscript C to Lewis 1980b, in Lewis 1986a, 125.
[34] Postscript B to Lewis 1980b, in Lewis 1986a, 120.

Lewis cites work by Brian Skyrms and Richard Jeffrey[35] who develop a clear conception of what he takes to be counterfeit chance. His own account of chance, he observes, can be seen as a special case of the Skyrms–Jeffrey notion of "objectified and resilient credence." The point is this: on one way of formulating Lewis's account of chance we first define a time-dependent partition of the possible worlds, where each cell of the partition consists of possible worlds that share the same history up to time t. Second, we define *a complete theory of chance*, T_w for a given world *w* as a function that specifies the chance, $P_{tw}(X)$, of each proposition X at each time *t* in world *w*. In this formulation, the Principal Principle says that for any reasonable initial credence function, C, $P_{tw}(X) = C(A/H_{tw}T_w)$. Lewis observes that this partition will satisfy three conditions: first, it is natural, rather than gerrymandered; second, "it is to some extent feasible to investigate (before the time in question) which cell of the partition is the true cell" and third, "it is unfeasible (before the time in question) . . . to investigate the truth of propositions that divide the cells." He then notes that there may also be coarser partitions that satisfy these three conditions, and he claims that it is reasonable to think that these assumptions are sufficient to ensure that if we have a reasonable initial credence function, then with enough "feasible investigation" our credences in events in the future (that divide the partition cells) would stabilize (or in Skyrms's terms, become resilient). Lewis's suggestion is that credences with this kind of potential to stabilize give rise to the appearance of unknown but discoverable objective probabilities, but he also cautions that any application of a more general account of "objectified and resilient credence" will depend on pinning down the vague notions of feasibility, and of naturalness, and for this reason will be "quite unlike genuine chance."

Whether these assumptions about the partitions are enough to ensure any kind of convergence may depend on what a priori constraints one puts on a reasonable credence function. De Finetti's theory gave us a precise argument for convergence, but it depends on the exchangeability condition, which, as I have suggested, can be motivated only with the help of causal assumptions. Still, I think that feasibility and unfeasibility conditions of the kind that Lewis gives are a part of what grounds a notion of objective chance, and Mellor's account of propensity helps to explain how they do. First, a concept of propensity, like any concept of a dispositional property, will have application only if there are ways of identifying things that have that property independently of whether the thing displays the manifestation property when subjected to the test condition. We don't need decisive criteria, but just the feasibility of evidence for it (this corresponds to Lewis's positive feasibility condition.) But second, Mellor's account implies that something will have a propensity with value *r* only if knowledge that it has it justifies you in having credence *r*, *whatever else you might know*. "Whatever else you might know" means whatever else it is feasible for you to know, which implies that if something has the propensity, it will not be feasible to know before the object is put to test, more about how it behaves than what is implied by the fact that the object has the propensity. That is, it must be that evidence or supposition that it has the propensity screens off all other

[35] Jeffrey 1983 and Skyrms 1977 and 1980.

information about its display. This condition is a causal independence condition: knowing more is ruled out because knowledge requires causal sensitivity to conditions that imply what is known. (This upshot of the propensity definition corresponds to Lewis's negative feasibility condition—that it is not feasible that one acquires evidence that divides the partition cell.)

I agree with Lewis that feasibility is a matter of degree, and may be more superficial, depending on practical limitations in the cognitive resources of particular knowers, or deeper, depending on limitations imposed by fundamental physical theory that make it empirically impossible for any cognitive being to be in states that are both sensitive to certain kinds of information and capable of determining that being's behavior. Indeterminacy in fundamental physics, as Lewis says, will be sufficient for the deepest kind of unfeasibility, but a deterministic theory can also allow for deep limitations in the possibility of the transmission of information. La Place hypothesized an omniscient demon who could predict the whole future in detail, but such a demon is no more physically possible than an oracle in an indeterministic world who sees directly into the future.

For the propensity theorist, the notion of chance is conceptually connected to assumptions about the possibility of knowledge, both constraining and being constrained by them. Knowledge depends essentially on causal relations between the knower and the facts known, but (I have argued) it is also a context-dependent notion, with truth-conditions for knowledge claims depending on presuppositions about normal conditions for the transmission of information.[36] The concept of chance, and other concepts in the natural necessity family will inherit some of the context-dependence of epistemic principles, but I don't think this is a threat to realism about these concepts, which requires only that they succeed in picking out genuine features of the world. In any case, a defender of Humean supervenience, and of the reduction of chance and other causal notions to the Humean base, is in no position to complain about indeterminacy and relativity when his most fundamental notions such as chance and causal dependence are explained in terms of laws of nature that are defined as generalizations that meet an unanalyzed standard that balances strength and simplicity, and comparative similarity relations between possible worlds that are defined in terms of vaguely specified priorities and measures of the size of deviations, in one world, from the laws of nature that hold in another. Lewis granted that it may be indeterminate what the best system of laws is, and exactly what comparative similarity relation between worlds is the right one for the interpretation of counterfactuals, but he did not take this to be a threat to his realist/reductionist project.

Skepticism based on chance

The third issue concerns a skeptical worry. "We think we have a lot of substantial knowledge about the future. But contemporary wisdom has it that indeterminism prevails in a way that just about any proposition about the future has a non-zero

[36] See Chapter 8 of this book.

objective chance of being false."[37] Can the hypothesis that the world is indeterministic in this way be reconciled with the possibility of substantial knowledge about the future? I have already suggested, in the discussion of epistemic undermining just above, that one might take oneself to know that a certain possible sequence of events with a nonzero chance of happening would not happen. (It is implicit in the example discussed that Alice took herself to know, before the start of a sequence of random draws of balls from a bag, that the first hundred draws would not all be red.) In that example, the sequence of events actually happens, so Alice did not know what she took herself to know, but suppose instead that things develop in the expected way, with black and red balls drawn about equally often, and in no discernible pattern. Then it seems reasonable to say that Alice really did know what she took herself to know, despite the fact that it had a non-zero chance of being false. How is this compatible with an account of knowledge that ϕ that requires the knower to be in an internal state that carries the information that ϕ? What the information-theoretic account requires is that the knower be in an internal state that, *under normal conditions* carries the information that ϕ, and that conditions in fact be normal (with the relevant normal conditions determined by the context of attribution of knowledge). In Alice's case, the relevant normal condition is what I have elsewhere called a "no misleading coincidence" condition. Conditions are abnormal in a relevant respect if there is evidence that is in fact coincidental but that would be highly probable on a salient alternative hypothesis. This is vague, and I don't try to make it more precise, but I hope the rough idea is clear, and that it is plausible.

In the case of the kind of highly improbable events that Hawthorne and Lasonen-Arnio have in mind in raising this skeptical worry (I drop a marble, and instead of bouncing and remaining on the floor, it quantum-tunnels through, disappearing from view), I think the general story should go something like this: Even if the true fundamental theory allows for this kind of event, there will be theories that are less fundamental, but true for the most part, and otherwise very much simpler. If the exceptions to the approximate theory are sufficiently improbable, we lose little by ignoring them. One way to ignore them in our cognitive behavior is to treat them as possible, but abnormal, and so not relevant alternatives. Of course, if the anomalous circumstances actually obtain in the situation, then the agent does not know what she reasonably takes herself to know, but in situations that are in fact normal in this sense, the agent will have the knowledge.

On the general picture I have been promoting, there are constitutive conceptual connections between, on the one hand, our concept of knowledge and the methods we develop and use to acquire knowledge, and on the other hand the theoretical resources we develop and use to describe the world. The aim is to reconcile this thesis with a realist view of theories stated with descriptive resources developed in this way. It is important for this reconciliation that the conceptual connections between epistemic rules and descriptive resources allow for the possibility of a gap between what the theories that are justified by our best epistemic resource tell us the world is like and what the world is really like. That is, the conceptual

[37] Hawthorne and Lasonen-Arnio 2009, 92.

connections do not ensure that skeptical hypotheses will not be true: we might be unlucky. The presupposed normal conditions that are essential to a characterization of our knowledge of the world are conditions that might in fact be false, in which case we don't have the knowledge we reasonably take ourselves to have. That is, even pervasive, undiscoverable deviations from normality remain conceptually possible. More specifically, the propensity theory's explanation of chance in terms of its relation to reasonable credence helps to ground the measurement of a quantitative property of things in the world, while also helping to constrain the credences that it is reasonable to have, but it allows for the possibilities of deceptive coincidences that persist, however much we are able to learn.

References

Adams, E. 1966. 'Probability and the logic of conditionals,' in *Aspects of Inductive Logic*, ed. by J. Hintikka and P. Suppes. Amsterdam: North-Holland, 265–316.

Adams, R. 1979. 'Primitive thisness and primitive identity,' *Journal of Philosophy* **76**: 5–26.

Ayer, A. 1936. *Language, Truth and Logic*. London: Victor Gollancz Ltd.

Ayer, A. 1972. *Probability and Evidence*. New York: Columbia University Press.

Battigalli, P. and G. Bonanno 1999. 'Recent results on belief, knowledge and the epistemic foundations of game theory,' *Research in Economics* **53**: 149–225.

Bennett, J. 2003. *A Philosophical Guide to Conditionals*. Oxford: Clarendon Press.

Berker, S. 2008. 'Luminosity regained,' *Philosophers' Imprint* **8**(2): 1–22.

Blackburn, S. 1980. 'Opinions and chances,' in *Prospects for Pragmatism*, ed. by D. Mellor. Cambridge: Cambridge University Press. Reprinted in Blackburn 1993, 75–93. (Page numbers to the latter)

Blackburn, S. 1993. *Quasi-Realism*. New York: Oxford University Press.

Block, E. 2008. 'Indicative conditionals in context,' *Mind*, **117**: 783–94.

Blome-Tillmann, M. 2009. 'Knowledge and presuppositions,' *Mind* **118**: 241–94.

Boër, S. and W. Lycan 1986. *Knowing Who*. Cambridge, MA: The MIT Press.

Boutilier, C. 1996. 'Iterated revision and minimal change of conditional beliefs,' *Journal of Philosophical Logic* **25**: 262–305.

Brandom, R. 1983. 'Asserting,' *Nous* **17**: 637–50.

Burge, T. 1988. 'Individualism and self-knowledge,' *Journal of Philosophy* **85**: 649–63.

Burks, A. 1951. 'The logic of causal propositions,' *Mind* **60**: 363–82.

Cappelen, H. and J. Dever 2013. *The Inessential Indexical: On the Philosophical Insignificance of Perspective and the First Person*. Oxford: Oxford University Press.

Carnap, R. 1936. 'Testability and meaning,' *Philosophy of Science* **3**: 419–71.

Chisholm, R. 1946. 'The contrary-to-fact conditional,' *Mind* **55**: 289–307.

Christensen, D. 1991. 'Clever bookies and coherent beliefs,' *Philosophical Review* **100**: 229–47.

Christensen, D. 2010. 'Rational reflection,' *Philosophical Perspectives* **24**: 121–40.

Cramér, H. 1955. *The Elements of Probability Theory and Some of its Applications*. New York: Wiley.

Darwiche, A. and J. Pearl 1997. 'On the logic of iterated belief revision,' *Artificial Intelligence* **89**: 1–29.

DeRose, K. 2004. 'Single scoreboard semantics,' *Philosophical Studies* **119**: 1–21.

Dorr, C., J. Goodman, and J. Hawthorne 2014. 'Knowing against the odds,' *Philosophical Studies* **170**: 277–87.

Dretske, F. 1981. *Knowledge and the Flow of Information*. Cambridge, MA: MIT Press.

Dretske, F. 2014. 'The case against closure,' in *Contemporary Debates in Epistemology* (second edition), ed. by M. Steup, J. Turri, and E. Sosa. Oxford: Wiley Blackwell, 27–40.

Dummett, M. 1975. "Wang's paradox," *Synthese* **30**: 201–32.

Edgington, D. 1986. "Do conditionals have truth conditions?" *Critica* **18**: 3–30. Reprinted in F. Jackson (ed.), *Conditionals*, ed. by F. Jackson. Oxford University Press, 1991, 176–201. Page references to the reprinted version.

Edgington, D. 1995. 'On conditionals,' *Mind* **104**: 235–329.

Edgington, D. 2004. "Counterfactuals and the benefit of hindsight," in *Cause and Chance; Causation in an Indeterministic World*, ed. by P. Dowe and P. Nordhof. London and New York: Routledge.

Elga, A. 2000. 'Self-locating belief and the Sleeping Beauty problem,' *Analysis* **60**: 143–7.
Elga, A. 2013. 'The puzzle of the unmarked clock and the new rational reflection principle,' *Philosophical Studies* **164**: 127–59.
Evans, G. 1979. "Reference and contingency," *The Monist* **62**: 161–89.
Fagin, R., J. Halpern, Y. Moses, and M. Vardi 1995. *Reasoning about Knowledge*. Cambridge, MA: MIT Press.
Field, H. 2001. *Truth and the Absence of Fact*. Oxford: Clarendon Press.
Field, H. 2009. 'Epistemology without metaphysics,' *Philosophical Studies* **143**: 249–90.
Fine, K. 1975. 'Review of Lewis, 1973,' *Mind* **84**: 451–58.
de Finetti, B. 1964. 'Foresight: its logical laws, its subjective sources' (translation of an article, originally published in 1937 in French), in *Studies in Subjective Probability*, ed. by H. E. Kyburg and H. E. Smokler. New York: Wiley.
Friedman, N. and J. Halpern 1999. 'Belief revision: a critique,' *Journal of Logic, Language, and Information* **8**: 401–20.
Gajewski, J. et al. 2003. *New Work on Modality* (MIT Working Papers in Linguistics and Philosophy), **51**.
Gärdenfors, P. 1988. *Knowledge in Flux: Modeling the Dynamics of Epistemic States*. Cambridge, MA: MIT Press.
Gärdenfors, P. and H. Rott 1995. 'Belief revision,' in *Handbook of Logic in Artificial Intelligence and Logic Programming* **IV**: *Epistemic and Temporal Reasoning*, ed. by D. Gabbay et al. Oxford: Oxford University Press, 35–132.
Gettier, E. 1963. 'Is justified true belief knowledge?' *Analysis* **6**: 121–23.
Gibbard, A. 1981. 'Two recent theories of conditionals,' in *Ifs: Conditionals, Belief, Decision, Chance and Time*, ed. by W. Harper, R. Stalnaker, and G. Pearce. Dordrecht: D. Reidel, 211–47.
Gibbard, A. 1990. *Wise Choices, Apt Feelings*. Cambridge, MA: Harvard University Press.
Gibbard, A. 2003. *Thinking How to Live*. Cambridge, MA: Harvard University Press.
Gibbard, A. 2012. *Meaning and Normativity*. Oxford: Oxford University Press.
Ginsberg, M. 1986. 'Counterfactuals,' *Artificial Intelligence* **30**: 35–79.
Goodman, J. and B. Salow 2018. 'Taking a chance on KK,' *Philosophical Studies* **175**: 183–96.
Goodman, N. 1983. *Fact, Fiction and Forecast*. Cambridge, MA: Harvard University Press.
Graff-Fara, D. 2000. 'Shifting sands: an interest-relative theory of vagueness,' *Philosophical Topics* **28**: 45–81.
Greco, D. 2014. 'Could KK be okay?' *Journal of Philosophy* **111**: 169–97.
Greco, D. 2017. 'Cognitive mobile homes,' *Mind* **126**: 93–121.
Grice, P. 1989. *Studies in the Way of Words*. Cambridge, MA: Harvard University Press.
Grove, A. 1988. 'Two modeling for theory change,' *Journal of Philosophical Logic* **17**: 157–70.
Hajek, A. 2010. 'Staying regular' (unpublished ms.).
Hajek, A. and N. Hall 1994. 'The hypothesis of the conditional construal of conditional probability,' in *Probability and Conditionals: Belief Revision and Rational Decision*, ed. by E. Eells and B. Skyrms. Cambridge: Cambridge University Press, 75–111.
Hall, N. 2004. 'Two mistakes about credence and chance, in *Lewisian Themes: The Philosophy of David K. Lewis*, ed. by F. Jackson and G. Priest. New York: Oxford University Press, 94–112.
Hawthorne, J. and M. Lasonen-Aarnio 2009. 'Knowledge and objective chance,' in *Williamson on Knowledge*, ed. by P. Greenough and D. Prichard. Oxford: Oxford University Press, 92–108.
Hempel, C. 1966. 'Recent problems of induction,' in *Mind and Cosmos*, ed. by R. G. Colodny. Pittsburg: University of Pittsburgh Press, 112–34.
Hild, M. 1998. 'Auto-epistemology and updating,' *Philosophical Studies* **92**: 321–61.

Hintikka, J. 1962. *Knowledge and Belief*. Ithaca: Cornell University Press.

Holliday, W. 2015. 'Epistemic closure and epistemic logic I: relevant alternatives and subjunctivism,' *Journal of Philosophical Logic* **44**: 1–62.

Horwich, P. 1990. *Truth*. New York: Oxford University Press.

Horwich, P. 1998. *Meaning*. Oxford: Clarendon Press.

Hume, D. 1748. *An Enquiry Concerning Human Understanding*.

Hume, D. 1896. *A Treatise of Human Nature*. Oxford: Clarendon Press.

Iatridou, S. 2000. 'The grammatical ingredients of counterfactuality,' *Linguistic Inquiry* **31**: 231–70.

Jeffrey, R. 1964. 'If,' *Journal of Philosophy*, **61**: 702–3.

Jeffrey, R. 1983. *The Logic of Decision* (second edition). Chicago: University of Chicago Press.

Kaplan, D. 1989. 'Demonstratives,' in *Themes from Kaplan*, ed.by J. Almog, J. Perry, and H. Wettsetein. New York and Oxford: Oxford University Press.

Kaufmann, S. 2004. 'Conditioning against the grain,' *Journal of Philosophical Logic* **33**: 583–606.

Kölbel, M. 2002. *Truth Without Objectivity*. London and New York: Routledge.

Kretzmann, N. 1966. 'Omniscience and immutability,' *Journal of Philosophy* **63**: 409–21.

Kripke, S. 1963. 'Semantical analysis of modal logic, I,' *Zeitschrift für mathematische logic und grundlagen der mathematic* **9**: 67–96.

Lehmann, D. 1995. 'Belief revision, revised.' *Proceedings of the Fourteenth International Joint Conference on Artificial Intelligence*, 1534–40.

Lehrer, K. and Paxson, T. 1969. 'Knowledge: undefeated justified true belief,' *The Journal of Philosophy* **66**: 225–37.

Lenzen, W. 1978. 'Recent work in epistemic logic,' *Acta Philosophica Fennica* **30**: 1–219. Amsterdam: North-Holland.

Lewis, D. 1968. *Convention*.

Lewis, D. 1973. *Counterfactuals*. Cambridge, MA: Harvard University Press.

Lewis, D. 1975. 'A problem about permission,' in *Essays in Honour of Jaakko Hintikka*, ed. by E. Saarinen et al. Dordrecht: Reidel, 163–75.

Lewis, D. 1976. 'Probabilities of conditionals and conditional probabilities,' *Philosophical Review*, **85**: 297–315.

Lewis, D. 1979a. 'Attitudes de dicto and de se,' *Philosophical Review* **88**: 513–43.

Lewis, D. 1979b. 'Scorekeeping in a language game,' *Journal of Philosophical Logic* **8**: 339–59.

Lewis, D. 1979c. 'Counterfactual dependence and time's arrow,' *Nous* **13**: 145–52. Reprinted in Lewis, 1986a, 32–86. Page references to the latter.

Lewis, D. 1980a. 'Index, context, and content,' in *Philosophy and Grammar*, ed. by S. Kanger and S. Öhman. Amsterdam: Reidel, 79–100.

Lewis, D. 1980b. 'A subjectivist's guide to objective chance,' in *Studies in Inductive Logic and Probability* 2, ed. by R. Jeffrey. Berkeley: University of California Press, 263–93. Reprinted in Lewis 1986a, 83–132. Page references to the latter.

Lewis, D. 1983a. 'Individuation by acquaintance and by stipulation,' *Philosophical Review* **92**: 3–32.

Lewis, D. 1983b. 'New work for a theory of universals,' *Australasian Journal of Philosophy* **61**: 343–77. Reprinted in Lewis 1999: 8–55.

Lewis, D. 1986a. *Philosophical Papers, Volume II*. Oxford: Oxford University Press.

Lewis, D. 1986b. *On the Plurality of Worlds*. Oxford: Basil Blackwell.

Lewis, D. 1996. 'Elusive knowledge,' *Australasian Journal of Philosophy* **74**: 549–997. Reprinted in Lewis 1999. Page references to the latter.

Lewis, D. 1999. *Papers in Metaphysics and Epistemology*. Cambridge, UK: Cambridge University Press.

MacFarlane, J. 2003. 'Future contingents and relative truth,' *Philosophical Quarterly* **53**: 321–36.
Mackie, J. 1973. *Truth, Probability and Paradox*. Oxford: Clarendon Press.
Martin, C. 1994. 'Dispositions and conditionals,' *The Philosophical Quarterly* **44**: 143–58.
Mayo, B. 1957. 'Conditional statements,' *The Philosophical Review* **66**: 291–303.
McGee, V. 2000. 'To tell the truth about conditionals,' *Analysis* **60**: 107–11.
Meacham, C. 2016. 'Ur-priors, conditionalization, and ur-prior conditionalization,' *Ergo* 3.
Mellor, D. 1971. *The Matter of Chance*. Cambridge, UK: Cambridge University Press.
Mellor, D. 1974. 'In defense of dispositions,' *Philosophical Review* **83**: 157–81.
Moss, S. 2012. 'Updating as communication,' *Philosophy and Phenomenologial Research* **85**: 225–48.
Moss, S. 2018. *Probabilistic Knowledge*. Oxford: Oxford University Press.
Nagel, E. 1961. *Structure of Science*. New York: Harcourt Brace.
Nagel, T. 1986. *The View from Nowhere*. New York: Oxford University Press.
Nozick, R. 1981. *Philosophical Explanations*. Cambridge, MA: Harvard University Press.
Osborne, M. and A. Rubinstein. 1994. *A Course in Game Theory*. Cambridge, MA: The MIT Press.
Perry, J. 1977. 'Frege on demonstratives,' *Philosophical Review* **86**: 474–97.
Perry, J. 1979. 'The essential indexical,' *Noûs* **13**: 3–21.
Putnam, H. 1963. '"Degrees of confirmation" and inductive logic,' in *The Philosophy of Rudolf Carnap*, ed. by P. Schilpp. LaSalle, Ill: Open Court, 761–83.
Quine, W. 1959. *Methods of Logic*. New York: Holt, Reinhart and Winston.
Quine, W. 1960. *Word and Object*. Cambridge, MA: The MIT Press.
Ramsey, F. 1929. 'General propositions and causality'. Published posthumously in F. Ramsey, *Foundations of Mathematics and other Logical Essays*, ed. by R. Braithwaite (1931). London: Routledge and Kegan Paul, 237–55.
Rescher, N. 1964. *Hypothetical Reasoning*. Amsterdam: North-Holland.
Rott, H. 1999. 'Coherence and conservativism in the dynamics of belief,' *Erkenntnis* **50**: 387–412.
Rott, H. 2001. *Change, Choice and Inference*, Oxford Logic Guides, 42. Oxford: Clarendon Press.
Rott, H. 2004. 'A counterexample to six fundamental principles of belief formation,' *Synthese* **139b**: 225–40.
Salow, B. 2016. 'Lewis on iterated knowledge,' *Philosophical Studies* **173**: 1571–90.
Schaffer, J. 2007. 'Knowing the answer,' *Philosophy and Phenomenological Research* **75**: 383–403.
Schroeder, M. 2008. *Being For: Evaluating the Semantic Program of Expressivism*. Oxford: Oxford University Press.
Schwarz, G. and M. Truszczynski 1992. 'Modal logic S4F and the minimal knowledge paradigm.' Proceedings of the Fourth Conference on Theoretical Aspects of Reasoning about Knowledge. Sam Mateo, CA: Morgan Kaufmann Publishers, Inc., 184–98.
Sellars, W. 1956/1997. *Empiricism and the Philosophy of Mind*. Cambridge, MA: Harvard University Press.
Shoemaker, S. 1980. 'Causality and properties,' in *Time and Cause*, ed. by P. van Inwagen. Dordrecht: Reidel.
Shoemaker, S. 1994. 'Self-knowledge and "inner sense",' *Philosophy and Phenomenological Research* **54**: 249–314.
Skyrms, B. 1977. 'Reliliency, propensities, and causal necessity,' *Journal of Philosophy* **74**: 704–13.
Skyrms, B. 1980. *Causal Necessity*. New Haven: Yale University Press.

Skyrms, B. 1984. *Pragmatism and Empiricism*. New Haven: Yale University Press.

Spohn, W. 1988. 'Ordinal conditional functions: a dynamic theory of epistemic states,' in *Causation in Decision, Belief Change and Statistics*, **2**, ed. by W. Harper and B. Skyrms. Dordrecht: Reidel, 105–34.

Stalnaker, R. 1968. 'A theory of conditionals,' in *Studies in Logical Theory*, ed. by N. Rescher. Oxford: Blackwell, 98–112. (Reprinted as Chapter 9 of this book).

Stalnaker, R. 1981. 'Indexical Belief,' *Synthese* **49**: 129–51. (Reprinted in Stalnaker, 1999a, 130–149.)

Stalnaker, R. 1984. *Inquiry*. Cambridge, MA: MIT Press.

Stalnaker, R. 1988. 'Belief attribution and context,' in *Contents of Thought*, ed. by R. Grimm and D. Merrill. Tucson: University of Arizona Press, 140–56. (Reprinted in Stalnaker, 1999a, 150–166.)

Stalnaker, R. 1991. 'The problem of logical omniscience, I,' *Synthese* **89**: 425–40. (Reprinted in Stalnaker 1999a, 240–54.)

Stalnaker, R. 1993. 'Twin earth revisited.' *Proceedings of the Aristotelian Society*, 297–311. (Reprinted in Stalnaker 1999a, 241–54.)

Stalnaker, R. 1994. 'What is a non-monotonic consequence relation?' *Fundamenta Informaticae* **21**.

Stalnaker, R. 1996. 'Knowledge, belief and counterfactual reasoning in games,' *Economics and Philosophy* **12**: 133–62.

Stalnaker, R. 1999a. *Context and Content: Essays on Intentionality in Speech and Thought*. Oxford: Oxford University Press.

Stalnaker, R. 1999b. 'The Problem of Logical Omniscience II,' in Stalnaker 1999a, 255–73.

Stalnaker, R. 2006. 'On logics of knowledge and belief,' *Philosophical Studies* **120**: 169–99. (Reprinted as Chapter 1 of this book).

Stalnaker, R. 2008. *Our Knowledge of the Internal World*. Oxford: Oxford University Press.

Stalnaker, R. 2010a. 'Responses to Stoljar, Weatherson and Boghossian.' *Philosophical Studies*, 1–13.

Stalnaker, R. 2011a. 'The essential contextual,' in *Assertion: New Philosophical Essays*, ed. by J. Brown and H. Capplelan. Oxford University Press, 137–50.

Stalnaker, R. 2011b. 'Putting Sleeping Beauty to rest,' *Canadian Journal of Philosophy*.

Stalnaker, R. 2011c. 'Conditional propositions and conditional assertions,' in *Epistemic Modality*, ed. by A. Egan and B. Weatherson. Oxford: Oxford University Press, 227–48 (reprinted as Chapter 10 of this book).

Stalnaker, R. 2012. *Mere Possibilities: Foundations of Modal Semantics*. Princeton, NJ: Princeton University Press.

Stalnaker, R. 2014. *Context*. Oxford: Oxford University Press.

Stalnaker, R. 2015. 'Luminosity and the KK thesis,' in *Externalism Self-knowledge and Skepticism*, ed. by S. Goldberg. Cambridge: Cambridge University Press, 17–40. (Reprinted as chapter 2 of this book).

Stalnaker, R. and R. Jeffrey 1994. "Conditionals as random variables," in *Probability and Conditionals: Belief Revision and Rational Decision*, ed. by E. Eells and B. Skyrms. Cambridge: Cambridge University Press, 31–46.

Stalnaker, R. and R. Thomason 1970. 'A semantic analysis of conditional logic,' *Theoria* **36**: 23–42.

Stampe, D. 1977. 'Toward a causal theory of linguistic representation,' *Midwest Studies in Philosophy* **2**, *Studies in the Philosophy of Language*, 43–63.

Stevenson, C. 1937. 'The emotive theory of ethical terms,' *Mind* **46**: 14–31.

Swain, M. 1974. 'Epistemic defeasibility,' *The American Philosophical Quarterly* **11**: 15–25.

Swanson, E. 2004. 'A suppositional theory of conditionals' (ms.).

Swanson, E. 2016. 'The application of constraint semantics to the language of subjective uncertainty,' *Journal of Philosophical Logic* **45**: 121–46.

Talbott, W. 1991. 'Two principles of Bayesian epistemology,' *Philosophical Studies* **62**: 135–50.

Thomason, R. 1970. "Indeterminist time and truth-value gaps," *Theoria* **36**: 264–84.

Thomason, R. and A. Gupta 1980. 'A theory of conditionals in the context of branching time,' *Philosophical Review* **89**: 65–90.

Thomson, J. 1990. 'In defense of ⊃,' *The Journal of Philosophy* **87**: 57–70.

Tolkien, J. 1966. 'On faery stories,' in *The Tolkien Reader*, by J. Tolkien. Ballantine Books: New York, 61–64.

van Fraassen, B. 1966. 'Singular terms, truth value gaps, and free logic,' *Journal of Philosophy* **63**: 481–95.

van Fraassen, B. 1976. 'Probabilities of conditionals,' in *Foundations of Probability Theory, Statistical Inference, and Statistical Theories of Science*, v. 1, ed. by W. Harper and C. Hooker. Springer, 261–308.

van Fraassen, B. 1984. 'Belief and the will,' *Journal of Philosophy* **81**: 235–56.

van Fraassen, B. 1995. 'Belief and the problem of Ulysses and the Sirens,' *Philosophical Studies* 77: 7–37.

Vogel, J. 1999. 'The new relevant alternatives theory,' *Noûs* **33**: 155–80.

Vogel, J. 2007.' Subjunctivitis,' *Philosophical Studies* **134**(1): 73–88.

von Wright, G. 1957. *Logical Studies*. New York: The Humanities Press.

Weber, C. 2015. 'Indexical beliefs and communication: against Stalnaker on self-location,' *Philosophy and Phenomenological Research* **90**: 640–663.

Weiner, M. 2005. 'Must we know what we say?' *Philosophical Review* **114**: 227–51.

Weisberg, J. 2007. 'Conditionalization, reflection, and self-knowledge,' *Philosophical Studies* **135**(2): 179–97.

White, R. 2005. 'Epistemic permissiveness,' *Philosophical Perspectives* **19**: 445–59.

Williamson, T. 1996. 'Knowing and asserting,' *Philosophical Review* **105**: 489–523.

Williamson, T. 2000. *Knowledge and Its Limits*. Oxford: Oxford University Press.

Williamson, T. 2011 'Improbable knowledge,' in *Evidentialism and its Discontents*, ed. by T. Dougherty. Oxford: Oxford University Press, 147–64.

Yablo, S. 2006. 'Non-catastrophic presupposition failure,' in *Content and Modality,* ed. by A. Byrne and J. Thomson. Oxford: Oxford University Press.

Yalcin, S. 2008. Modality and Inquiry. PhD thesis, MIT, Department of Linguistics and Philosophy.

Yalcin, S. 2012. 'Bayesian expressivism.' *Proceedings of the Aristotelian Society* **112**: 123–60.

Index

The manufacturer's authorised representative in the EU for product safety is Oxford University Press España S.A. of El Parque Empresarial San Fernando de Henares, Avenida de Castilla, 2 - 28830 Madrid (www.oup.es/en or product.safety@oup.com). OUP España S.A. also acts as importer into Spain of products made by the manufacturer.
Printed and bound by CPI Group (UK) Ltd, Croydon, CR0 4YY
14/09/2026
02216288-0004